ACCLAIM FOR HENRY PETROSKI'S

SMALL THINGS CONSIDERED

"Fascinating. . . . [Petroski] has combined a writer's grace with an engineer's insight to give us an engaging series of essays. . . . You'll never again take a potato peeler for granted." —*St. Louis Post-Dispatch*

"[Petroski] shares with Carl Sagan, Stephen Jay Gould, and Stephen Hawking a talent for taking his passion and making it accessible to those who lack his scientific background while being sufficiently observant and meticulous to keep it interesting for those who share it." —*Civil Engineering*

"An engaging read." —*The Denver Post*

"Fascinating. . . . Interesting and insightful observations. . . . Petroski will make any reader . . . more aware of the processes that lead to the variety of things that are all around us and how they came to be the way they are." —*Science Books & Film*

"Craftily, [Petroski] combines an engineer's insight and admiration for the way things are designed with a layman's puzzlement." —*Boston Herald*

HENRY PETROSKI

SMALL THINGS CONSIDERED

Henry Petroski is the Aleksandar S. Vesic Professor of Civil Engineering and a professor of history at Duke University. He is the author of ten previous books.

Also by Henry Petroski

Paperboy: Confessions of a Future Engineer

The Book on the Bookshelf

Remaking the World: Adventures in Engineering

Invention by Design: How Engineers Get from Thought to Thing

Engineers of Dreams: Great Bridge Builders and the Spanning of America

Design Paradigms: Case Histories of Error and Judgment in Engineering

The Evolution of Useful Things

The Pencil: A History of Design and Circumstance

Beyond Engineering: Essays and Other Attempts to Figure Without Equations

To Engineer Is Human: The Role of Failure in Successful Design

SMALL THINGS CONSIDERED

SMALL THINGS CONSIDERED

WHY THERE IS NO PERFECT DESIGN

HENRY PETROSKI

Vintage Books

A Division of Random House, Inc.

New York

FIRST VINTAGE BOOKS EDITION, SEPTEMBER 2004

 Published in the United States by Vintage Books, a division of Random House, Inc., New York, and simultaneously in Canada by Random House of Canada Limited, Toronto. Originally published in hardcover in the United States by Alfred A. Knopf, a division of Random House, Inc., New York, in 2003.

The Library of Congress has cataloged the Knopf edition as follows:
Petroski, Henry.
Small things considered: why there is no perfect design / Henry Petroski.
p. cm.
1. Engineering design. I. Title.
TA174.P4738 2003
620'.0042—dc21
2002192470

Vintage ISBN: 978-1-4000-3293-8

Author photograph © Catherine Petroski
Book design by Robert C. Olsson

www.vintagebooks.com

146028962

to Catherine

CONTENTS

SMALL THINGS CONSIDERED

ONE

The Nature of Design

An architect, a psychologist, and an engineer were appearing together on a radio talk show. This may sound like the setup line for one of those jokes about professionals, but the situation was real and the subject of the National Public Radio program, *Talk of the Nation—Science Friday,* was not comedy but the design of everyday objects. The hour's discussion showed that the participants, each in his own way, took design very seriously indeed. Their characteristic mind-sets were evident from the beginning, and by the end of the hour they had conveyed a great deal about how differently their professions approach the subject under discussion.

Ira Flatow, the show's host, asked each guest to comment on what constituted good and bad design—giving examples—and the talk ranged widely from teakettles to computers. As if to signal that this was not too heavy a subject for a Friday afternoon, the architect stressed at the outset that the design of anything can be whimsical and a source of great enjoyment. He recalled getting a letter from a French poet who said that he smiled every time the little red bird on his Michael Graves–designed teakettle whistled to announce that the water had come to a boil. The psychologist agreed that the architect's teakettle was a thing of beauty and a joy to use, and he confessed that he even displayed one in his kitchen window for people to admire. The engineer, having had no personal experience with the whimsical teakettle, said nothing about it, but he thought to himself of the many times he had dripped boiling water from the very attractive but poorly designed kettle that sat on his stove at home.

Throughout the show, the architect spoke often of aesthetics and about how much of design should be driven by the appearance of things, if for no other reason than to distinguish from one another in the marketplace objects that serve the same purpose. The psychologist, while acknowledging that he liked attractive designs, stressed how important it was that an object communicate how it was to be used. Among the examples of poor "user interfaces" he cited in the course of the show were feature-laden telephone systems and shower controls that gave no hint of how they were operated.

The engineer did not disagree with the architect and psychologist. He, too, expressed a wish that the design of objects be attractive enough that they could be displayed in a place of honor and their use be transparent. However, he pointed out, the primary purpose of most things is to perform a function, and because the goals of aesthetics, user friendliness, and doing a job effectively can be in conflict, economics often becomes the referee. The design process is characterized mostly by tensions between competing objectives that are resolved by compromises, usually driven by the realities of manufacturing cost and sales price. After some discussion, which led to a collegial agreement that good design must take into account, within the context of cost, all manner of considerations, the host invited the listening audience to call in and describe their favorite examples of good design and also relate their greatest "design nightmares."

Of the many things named by the host, the guests, and call-in listeners over the course of the show, there were about equal numbers of what were considered good and bad designs. Among the "good" designs were musical instruments, the Chrysler Building, and the paper clip. "Bad" designs included car-radio buttons, electronic remote controls, and water pitchers. But the dividing line between good and bad design was not drawn sharply, and opinion was not always unanimous. One caller offered the modern automobile steering wheel as an example of good design: By having so many controls at their fingertips, drivers do not have to take their eyes off the road to fiddle with knobs and buttons, thus making driving safer. Another caller criticized the modern steering wheel as a safety liability for not having, as older cars did, an obvious button or ring for beeping the horn. No sooner had one caller praised the fashionable fat-handled kitchen gadgets as a godsend for older people than another listener called to complain that they are not so wonderful for those, like her, who are allergic to latex. Reacting to the material on a new set of utensils given her as a gift, she removed the molded covering from the handles before she continued to use the things, leaving them unsightly and rather uncomfortable to hold. At the end of the program, the only clear conclusion was that it can be as hard for three professionals to agree on the essence of design as it is for three sightless men to agree on the nature of an elephant. But no matter what it is, design is certainly no joke.

During a different public radio show, on which the engineer was the lone guest, the question of the design of common objects arose once again. The host preempted the design cynics by asking listeners to discuss only good design, and she challenged the audience to call in with their favorite examples. One listener hesitatingly began describing an object that she was not sure anyone else would appreciate but which she marveled at when she had a hot pizza delivered: the "thingy" that keeps the top of the box from sagging and getting stuck to the melted cheese. Though she did not know what it was really called, the white plastic tripod she described was immediately recognizable to those in the studio and control room, all of whom smiled their assent.

The caller's enthusiasm for the object made it clear that she thought the pizza-box insert was the epitome of design. She told of washing and saving the little tripods, expecting someday to find new uses for them. The host of the show again nodded her silent approval. The engineer confessed to having saved some of the humble devices as examples of clever functional design that he thought he might write about someday. No one called in to ask for a better description of the throwaway thing, or to offer its name. But such identification was not needed for it to be recognized, admired for its ingenuity, and appreciated for its purpose. After the show aired, another listener, an artist, sent an E-mail message, saying that she also admired the design of the white plastic objects, which she called "triangle platforms." She described shortening their legs and using them as spacers between stacked palettes in her paint-storage box. She had also used them for a different purpose, turning them upside down to support, like little Atlases, spherical objects for display. Things are often used for purposes other than what their designers intended or even imagined.

Another fan of the plastic tripods finds them ideal for holding eggs, to which she applies sequins, beads, and other festive trim to make Christmas-tree decorations. For this admirer, the simple devices are definitely not throwaway items. In fact, on the illustrated Web site describing her utilization of these objects, they are lightheartedly evaluated:

> The plastic tripod is very expensive, costing somewhere around $10 US, but it is worth it for the ease of working with the eggs. It's prob-

Plastic tripods designed to keep the top of a pizza box from coming in contact with melted cheese also have other uses.

ably the packaging that makes the tripod so expensive. It comes packaged with a Pizza Hut carry out pizza. The pizza and the cardboard box protect the plastic tripod, though some would assert that the box and tripod protect the pizza. Regardless, the pizza and box can be discarded in some ecologically sound manner and the tripod used to hold the egg as a work in progress.

However viewed, the pizza-box platform presents a fine example of something designed to be very functional, but not something most of us would think to put on display in our windows—or something whose use would be at all apparent out of context. Yet at least one listener thought enough of the trivial little object to sing its praises over public radio, another to send a note about it, and a third to describe it on a Web site. If there were people listening to the radio program who sneered at the nomination of the pizza-box throwaway device for favorite-design honors, they did not bother to call or write.

No design, no matter how common or seemingly insignificant, is without its adamant critics as well as its ardent admirers. And no object is anonymous to the cognoscenti. In fact, the plastic tripod is sold as a "pizza saver" by the family business in Coatesville, Pennsylvania, that provides hotels and restaurants with this item, as well as with disposable paper products and other items, such as toothpicks and stirrers. The three-legged specialty item, offered in the Royal Paper Products catalog as a tripod to "keep the top of the pizza box from sticking to the cheese," is indeed a boon to fastidious pizza lovers, and its tiny feet barely disturb the toppings. As a designed object, however, it might be criticized for being too blandly functional and cheap looking. It is not difficult to imagine some critics objecting to its lack of style, others to the fact that it is made of plastic, and still others to the way strings of cheese stick to the spindly legs. Admirers of the little box top–propping scaffolds might retort by saying, "So, design something better." In the case of something like a pizza saver, most people will say, "Who needs something better? It works fine." But even this trivial little thing has undergone design changes, with the original Royal Paper model having been made of polypropylene and a later one of less expensive polystyrene.

Nothing is perfect, if by perfect one means absolutely free of every flaw or shortcoming, even if the drawback is that the seemingly insignificant cost of a packaging item is considered too expensive in the aggregate. Because every design must satisfy competing objectives, there necessarily has to be compromise among, if not the complete exclusion of, some of those objectives, in order to meet what are considered the more important of them. The pizza saver, like the typical flat box in which it is found, is functional but not especially beautiful. The squat tripod does the job it was designed to do, but if it were first encountered in a kitchen drawer, that job would not be obvious. The architect and psychologist were not around to rate the plastic pizza protector as an object of design, but they might have found it wanting because of its lack of styling and its failure to shout out its use. Judged against the standards of museums and upscale gadget emporiums, the tripod might not have a chance.

The economist Herbert Simon introduced the term *satisficing,* a combination of the words *satisfying* and *sufficing,* to refer to "good or satisfactory solutions instead of optimal ones" for design problems. According to Simon, the "decision maker has a choice between optimal decisions for an imaginary simplified world or decisions that are 'good enough,' that satisfice, for a world approximating the complex real one more closely." Design is nothing if not decision making. Someone somewhere at some time had to make decisions about what the pizza-box device would look like. In order for it to be made, someone had to decide on the number and shape and size of its legs and how they would be joined. Someone had to specify the material to be used, and its color and texture. The resulting design, which sufficiently satisfies the requirement of preventing the box top from coming into contact with the pizza, is certainly good enough for the modest role that it plays in the real world of real things. That the pizza-box tripod can also serve, albeit unintentionally, for holding round and ovoid objects for display and decoration just makes its design all the more satisficing. Indeed, though it may never garner awards for aesthetic excellence, admirers see it as a thing of beauty.

We do speak of "perfected" designs, of course, but in this context the word *perfect* is understood to have the dictionary definition of "having all the required or desirable elements, qualities, or characteristics" but not of being "as good as it is possible to get" given the existence of competing goals. How good something can possibly get is a nebulous concept, because we cannot know how good that is until something gets that good, and even then, can it ever be truly perfect? It's not just that we'll know good design when we see it; it's that we'll see it when we hold it in our hands, or sit in it, or jump up and down on it, or walk across it, or open up a pizza box and find it inside. In the meantime, what we do have must suffice until some improved thing comes along. And we do make things suffice, to the point of growing accustomed to them and rationalizing their shortcomings.

It is telling that the titles of so many patents, those official documents in which designers known as inventors reveal new ideas and things to the world, begin with the words "Improvement in." So much of invention and design is, in fact, but improvement on an earlier improvement, a refinement of the "prior art," as the unimproved technology is referred to in the patent literature. This is so because, nothing being perfect, there is always room for improvement. The concept of comparative improvement is embedded in the paradigm for invention, the better mousetrap. No inventor or designer is likely ever to lay claim to a "best mousetrap," for that would preclude the inventor herself from coming up with a still better mousetrap without suffering the embarrassment of having previously declared the search complete. There can never be an end to the quest for the perfect design.

Even in nature, absolute perfection, in the sense of something being completely flawless, is rare, if not totally absent. The close examination of any botanical or zoological specimen always reveals irregularities at some power of magnification—a tiny blemish or a mutant gene. Even in the mineral kingdom, where silicon presently reigns, purity is measured in parts per billion, implying that there are impure parts to count. Ore is never 100 percent fine, and no gem is without a flaw of some kind. The designation "perfect storm," popularized in recent years as the epitome of bad weather, when all unlikely circumstances come together in one place at one time, is, in fact, a pejorative to fishermen.

Artists might be said to be designers who can ignore user transparency and common function in producing "art for art's sake." Nevertheless, some artists might be said to seek perfection in their technique, rather than in their subjects. Andrew Wyeth did not avoid the torn curtain or the ordinary model. Neither did Picasso find perfection in his models, and his portraits emphasize the asymmetry of even the most beautiful face. His women with eyes at different levels and in different orientations are exaggerations but not misrepresentations of the geometry of the face as it is found. Indeed, the totally symmetrical face would appear not to belong to a human being at all, but to a computer-generated image. It would be "too perfect" to be real.

The design of made things, as opposed to design in nature and the artistic interpretation of it, necessarily proceeds within the confines of the laws of science and economics. An artist may paint a woman floating in the air on her birthday, and a collector may spend millions of dollars on a painting of sunflowers that he will hang in his private gallery. But an inventor or designer of practical things must accept the realities of gravity and budgets, keeping his feet on the ground and his eye on the price.

The design of anything in the real world must always be done within constraints, those (initially, at least) nonnegotiable limitations and restrictions that are imposed upon the designer by a budget, by a boss, by a client, or even by the designer herself. Some engineers even define the creative aspect of engineering as "design under constraint," to emphasize that what engineers do is always tied to the reality of the world and of the budget. Architects are also expected to work under such constraints, though, like all designers, they do not always manage to do so. The Sydney Opera House, though an icon of international renown, is famous for the impracticality of its initial design and for its final cost, which was fourteen times the designated budget.

Consider the design of something less grand, like a garage. The principal functional objective is simple: to provide an enclosed space in which to keep an automobile when it is not in use. If the garage is to be built adjacent to a house, it must conform to various constraints. It must be wholly within the property line; be respectful of setbacks;

be consistent with any zoning rules for size and height; be large enough to accommodate the family car; have a door large enough for the car to enter and exit; be located so that it is accessible via a driveway; be affordable and proportionate in cost. Such hard constraints would be nonnegotiable if the resulting structure were to function as a legitimate garage. In addition, there would be softer constraints, which might be ignored, but at the expense of an odd-looking garage. Such constraints might include that the structure be in a style similar to the house; be on a scale consistent with the house; be located beside or behind but not directly in front of the house; be painted in a color consistent with the house.

All design involves choice, and the choices often have to be made to satisfy competing constraints. Building a two-car attached garage beside a house that is only a single car width from the property line presents a design problem with seemingly incompatible constraints. There are solutions, but all involve compromise, which leads to a less than ideal solution. One would be to build the garage well behind the house, connected to it by a breezeway, but this would necessitate using considerable space in the backyard, where the driveway would have to be doubled in width, something that might be undesirable. Another solution would be to build a garage one car width wide but two car lengths deep. This tandem garage need not encroach upon the backyard at all, but it would necessitate moving one car whenever it blocked the exit of the other. The designer must always make choices within constraints, hard and soft.

Inevitably, realizing a design involves cost. There is, of course, the direct monetary cost of materials and labor, but there are also indirect costs associated with aesthetics, convenience, the environment, and other factors that are not strictly functional. These costs are seldom clear-cut, for many of them involve subjective judgments relating to things in the eye of the beholder or the politics of the opposition. Few neighbors would like a garage built right up to their property line or one that was, for reasons of cost or nastiness, not finished on the side at which they, but not the garage's owners, would have to look. Choice and cost are intimately connected in any design.

In the interest of maintaining good neighborly relations, most homeowners will make their garage narrower than they would like in order to allow themselves a path beside it that will not encroach on the neighbor's backyard. Most likely, they will also finish the side of the garage in a manner consistent with the front, not only to show the neighbor that they are sensitive to her concerns but also as a matter of aesthetic integrity for the design itself. In some cases, they may erect a fence beside the garage, to keep the neighbor's kids from playing ball against it.

Some people say, as Robert Frost did, that good fences make good neighbors. Not every fence is a design that a neighbor can live with, of course, but some are neighbor-friendly. A low post-and-rail fence, though a physical impediment, is most often innocuous and easily talked over or climbed through or over by pets, children, and even the neighbors themselves. It also looks pretty much the same from either side. A typical white picket fence is essentially a post-and-rail fence faced with uniform uprights on one side only. The simple structure can present a considerable design dilemma: Who will see the prettier, more finished, side?

When built around the front of a house, a picket fence usually has the horizontal rails facing in toward the home, so as to present to passersby and visitors a neat appearance to complement the house's facade. However, a picket fence built around a backyard located between two other backyards presents a different problem. As is true with the fence in the front yard, the principal constraint is that the pickets must face either toward or away from the backyard. Tradition and neighborly courtesy might dictate that the finished side face out here, too. But if the iconoclastic homeowner is so annoyed that neighbors across the street have a better view of his front fence than he does, he might consider installing his backyard fence with the rails facing his neighbors. He might see the fence's better side from his patio, but the arrangement clearly would impose on the neighbors a view that he does not wish for himself. The choice may or may not be difficult, depending on the personality of the fence builder. Unless he wants to incur the added expense of facing both sides of the rails with attractive pickets, an apparent compromise that itself might result in an even uglier fence, he will have to make a choice. (Another compromise is to alternate the side to which facing boards are attached, but this "good neighbor" solution is not common when constructing picket fences.) If he is not willing to compromise his view for that of his neighbors, then he necessarily factors into the design of his fence the cost of lost goodwill.

An angry neighbor who has had his aesthetic sensibilities offended by the erection of a picket fence whose rails face into his yard might retaliate with a fence of his own. A tall plank fence with its rails facing back into the first neighbor's yard would accomplish two things. First, it would hide from view the backward picket fence, as well as the neighbor who erected it. Second, it would give that neighbor a dose of his own medicine, showing him the back side of a fence, which would definitely diminish the effect he was seeking with his picket fence. Which of these in-your-face designs would result in the less neighborly fence would, no doubt, depend on one's point of view.

Designing anything, from a fence to a factory, involves satisfying constraints, making choices, containing costs, and accepting compromises. These givens necessarily introduce individual characteristics and anomalies into the resulting artifact. Constraints are typically a major part of the defining features of a design problem, and so they become dissolved and absorbed into the solution. Choices must be made among the conditions that satisfy the constraints. The necessity of compromises may be less evident at the outset, but they are an adjunct to choice and tend to characterize the solution. It is the rare case in which no compromise at all has to be made in form or function, and it is the nature of the compromises arising from the choices among the constraints that makes a design less than perfect. Does this mean that design is always a downer, its products failing to satisfy fully or perhaps even to suffice?

Clearly, there are admirable designs, elegant solutions to difficult problems, that celebrate the ability of inventors to hit a home run now and then. But just as in baseball, where, once, hitting five dozen home runs out of five hundred at bats amounted to a record-breaking season, expecting a homer every time is unrealistic. Just getting any kind of hit four times out of ten is a legendary performance, and designers, like sluggers, are judged more by their hits than by their misses. Designers know that not every hit will be a home run, or that every home run will be a grand slam, or even send the ball out of the park. When we recognize the difficulty of a human endeavor, and the odds against success in it, we judge performance by less than perfect standards.

Even though no design can be perfect, that does not mean that every design is a failure. We evaluate designs not against absolutes but against one another. The better mousetrap is just that—better—and this may be reason enough to admire it. Frequently, a design is praised merely for its aesthetics, with other constraints shadowed by the glow of the beautiful object. No matter that a strikingly handsome chair is uncomfortable to sit in or impossibly expensive to own. Though architects may prefer form over function, engineers may sacrifice aesthetics in meeting function, and so a sturdy bridge might blight a highway's natural beauty. Neither form nor function should overwhelm the other in design, but in the real world, one is often achieved at the expense of the other.

Few critics would disagree that in the ideal world all made objects should be designed to be as attractive, usable, and efficacious as possible, within the constraints of the defining problem. In fact, though, how well things look and work is often what governs how they are judged as designs. In particular, the user interface, though ideally transparent, usually gets short shrift compared to considerations of form and function. Indeed, Donald Norman, the psychologist–cum–computer scientist who has written thoughtfully about the importance of user interface in design, has admitted that if his advice were followed with absolute strictness, "things would be useful but ugly."

Beauty is ultimately in the eye of the beholder, of course. We tend to develop an affection and a fit for our familiar tools and furniture, no matter what they look like. Our favorite hammer or chair is often the one that we have grown so accustomed to that any other feels awkward or uncomfortable to use. Our favorite things get old and worn, but they become so molded to our shape that we do not care that they are dirty and deformed and possibly even offensive to the senses of others.

By absolute aesthetic standards, they may be downright ugly. But judged by personal aesthetics, they may belong in a museum. Judging design is as subjective as judging anything else, and it relies greatly on context.

The ultimate context of design is, of course, the human user. Many designed things are "one size fits all," and so if they fit anyone perfectly, it is a statistical coincidence. This being so, all the rest of us must make do. Sometimes we can shop around and try a different brand or model of a designed object, hoping to find the one that seems to have been made for us. Most likely, we never find such a thing, and so we compromise in our choice, selecting a less attractive chair because it is more comfortable or picking an uncomfortable chair because it looks more striking in our living room.

We learn to live in a world of imperfect things, just as we do in a world of imperfect fellow human beings. If we cannot find a pair of shoes that is a perfect fit for us, and if we cannot or do not wish to spend the money to have our shoes custom-made, then we choose a pair whose looks and fit are as close to what we want as we can find. We know from experience that there will be a breaking-in period, but if we choose wisely, in time the shoes will conform to our feet, or our feet to them. By the time the shoes are totally comfortable, they are usually either out of style or looking rather worn, but we have broken them like a cowboy breaks a horse. We have been victorious over the tyranny of design and the object that results.

We think, therefore we design. Indeed, there is barely anything that we do, much less use, that does not have a design component to it. How do we watch television? We arrange chairs before the television set, not behind it, and we orient them so that they face the screen. We select what program we want to watch, often in consultation with the others in the room. If there is not unanimity over what to watch, we compromise, at least in the ideal house. If we have rented a movie, we decide together when to stop the tape to take a bathroom break or make popcorn. We design an evening before the television set.

Because so much of what we do involves design of one kind or another, we know instinctively what design entails. We understand that we cannot watch a television show at 8:00 p.m. that does not air until 9:00. We understand that we cannot watch two programs at once, unless we have more than one set or our set has a picture-in-picture feature. We understand that, even if we have the technical capability, we cannot really watch two mystery movies simultaneously and expect to keep both plots straight. We understand the necessity of choice and the cost of trying to circumvent it. We understand that we cannot have the sound low enough that the sleeping baby will not hear it and loud enough that we can hear the show without straining our ears or missing some of the crucial dialogue. Because we understand the nature of so common an activity as watching television, we understand the nature of design, and we understand the elusiveness of perfection. That nothing is perfect is not an indictment of design, but an acknowledgment of its human origins.

Since we are all designers ourselves, even if only in arranging a television-watching evening, we do have some experience to render us creditable critics of design. Yet easy as it is to criticize a designed thing, improving it can be difficult, which is why inventors are rewarded with patents and the privileges that they carry. Since most of us do not wish to or have the inclination to go beyond criticism to formal invention, we accept things as they are, imperfect as they are, or we modify our use of them. We adapt and conquer, or at least make do.

Though there may be no perfect design, we can still speak of good design. We can admire the brilliant solution, appreciate the ingenious device, and enjoy the clever gadget. Imperfect as they may be, they represent the triumph of the human mind over the world of things, and the achievements of accomplished designers uplift the spirit of us all. The pole-vaulter who sets a new record is no less of a champion because he does not clear the next bar height. He had conceived and executed his run, the planting of his pole, and the arc of his body in the best way that he could for that meet, and for the time being, at least, his best is the best. We applaud what he did achieve, with the expectation that someday he or some other athlete may design a better pole or vaulting technique and so set a new record. That is the nature of design.

TWO

Looking at Design

I HAVE a glass of water on the table beside me. It is a medium-size glass, capable of holding about ten ounces of liquid when filled to the brim. The glass is clear, the same colorlessness as the water that it holds, though both the glass and the water do pick up the colors of items scattered about on the desk—the blue of the coaster, the yellow of the pencil, the red of the pen. A talented artist might make the glass of water the focus of an interesting still life, but the geometry of the scene might have challenged even the meticulous draftsman M. C. Escher. The way the things on the desk are reflected in and refracted through the water and the glass suggests to me the effect of a fun-house mirror, an impression reinforced by the curved surface of the glass. Though the body of the vessel itself is a simple circular cylinder, of about the usual thickness for a drinking glass, it has a thick, heavy bottom with a single large air bubble captured in the center of the base. When I lift the glass to drink from it, the bubble marks the vessel itself as something designed, as an objet d'art as well as a practical container for liquids. If I look more carefully at the bubble, I see that it acts like the crystal spheres that Escher held in his hand, capturing in the round, as does a fish-eye lens, the arrangement of the room surrounding it.

For all of its artistry, this glass is primarily an object of functional design. Its size is just about right: Its ten-ounce capacity means it holds a good amount of water without being so heavy as to be uncomfortable

The text on a page is distorted by a glass sitting on it.

to lift. Its three-inch diameter is about right for an adult hand to get a good solid grip on, and its squat four-inch height means that it is not easily tipped over by a hand or arm reaching for something beyond it. (This glass is not supposed to be a tumbler.) Its thick, heavy bottom adds to its stability, and the flat base allows the glass to sit solidly on my desk, without rattling as I type. Its round cylindrical shape and symmetry mean that I do not have to think about where around its periphery I should grasp the glass or from what place around the rim I should drink. The fact that the glass is clear makes it easy to tell whether it contains milk, water, or iced tea. The gentle filleting of its inside vertical surface into the bottom horizontal one eliminates hard-to-reach corners into which dried milk or orange juice could collect. The well-proportioned rounding of its rim and the outside edges of its base minimizes the chance of the glass being chipped in the dishwasher. Its straight vertical profile makes it easy to store compactly in the cupboard with other glasses of the same design.

Drinking glasses, though perhaps not as well proportioned, symmetrical, and artistically designed as this one, were invented (first designed) millennia ago. All modern glasses may be thought to be variations on an ancient theme. The glass on my desk might be said to be a perfected design, without implying that it has no faults and so cannot be improved upon. This glass might be just the right size for use at my desk, but I might want one with a larger capacity if I were drinking iced tea on the lawn in the heat of summer. I definitely would prefer a smaller glass in which to serve a liqueur or port after dinner. Glasses of different sizes and shapes are appropriate for different purposes and different occasions. This is why our cupboard holds three dozen glasses of three different sizes, each with a bubble in its base, and our china cabinet contains a miscellany of other glasses. It might easily be argued that we should have an even wider variety of sizes and types, but there are limitations to how much cupboard and cabinet space can be devoted to storing glasses, not to mention how much space the manufacturer and distributor might want to allow for making and stocking them. (A recent visit to the store where my wife and I bought our drinking glasses a few years ago confirmed that this Bubble glass, as the style is named by the manufacturer, does indeed come in only the three sizes that we have: the seven-ounce juice, the ten-ounce double old-fashioned, and the sixteen-ounce highball.)

The rounded rim on the ten-ounce Bubble glass holding water on my desk makes it comfortable to drink from and presents no sharp corners to my mouth. Though it might be difficult to argue for any functional advantage to a glass with severe corners, some hard-edge "designer" glasses do appear to be made that way. Design in the sense of aesthetics and fashion often goes counter to design in the sense of function, use, and common sense. I was once offered a drink in a glass of squarish shape. It looked chic when arriving on the serving tray, but it presented a challenge to drink from. A square glass is naturally grasped across its flat sides, but in doing so, the wrist has to be twisted to bring a corner of the glass to one's mouth. Otherwise, some of the liquid will flow past the lips, down the chin, and onto the shirt. Even with the square glass in the proper orientation, drinking from it has to be done with care, since the corner funnels the liquid toward the mouth at a greater rate than most of us are accustomed to in a conventional (round) glass raised at the same angle. In other words, the square glass is essentially (and, one hopes, unintentionally) a disguised dribble glass. A set of such glasses might store more compactly in the cupboard and look striking on the coffee table, but their storage and display should be secondary to how the glasses function. A glass in the hand is worth more than two with the dishes.

The round glass on my desk is empty now, and I do not have waiter service in my study. Since I did not bring a bottle of water with me this morning, if I want more water, I will have to stop working and take the glass to the kitchen to refill it. That is no big deal, but wouldn't it be nice if drinking glasses were self-refilling? And self-cleaning? Wouldn't it be nice if glasses did not have to be loaded into the dishwasher, then unloaded and put away in the cupboard? Wouldn't it be nice if attractive glasses automatically kept cold drinks cold (without sweating) and warm drinks warm (without burning the hand that holds them)? Wouldn't it be nice if the contents of glasses did not spill when they were tipped over? Wouldn't it be nice if glasses didn't break when dropped on a quarry-tile floor? Wouldn't it be nice if one glass served all purposes, and fit them all fashionably? Wouldn't it be nice if glasses were perfect?

Such are questions that drinkers might ask when using glasses. But inventors, engineers, and designers ask and answer such questions all the time about everything they see, touch, and use. One clearly does not have to be a professional critic of technology to ask such questions. We all recognize that the things we use every day are not perfect, and most of us can think of ways in which they could be improved. But thinking about an attractive glass that doesn't spill or one that doesn't ever need to be washed is a lot easier than making one. Nonspill glasses for babies are popular in the nursery, but what host or hostess would serve cocktails in the living room or wine at the dining table in brightly colored plastic tumblers with lids and spouts? And anything that would never have to be washed would seem intuitively to violate some law of nature. As much as we recognize that just about everything could be improved upon, we also realize that there are limitations to what can be done with the resources available to us. Among the most limiting of those resources are materials, money, and imagination.

All materials come ultimately from nature. Even so-called man-made materials are composed of atoms, the building blocks of the universe. The principal material from which glass is made is basically sand heated to a very high temperature, at which time it fuses into a liquid state and so can be formed to harden into objects like vases, bottles, and the drinking glass on my desk. (The embedded bubble in my glass is the result of a clever fillip, a seeming tour de force of the traditional glass-maker's art.) The fact that the name of the object itself—glass—echoes the material of which it is made suggests how basic materials are, especially in the most common of things. We also have window glass, looking glasses, magnifying glasses, eyeglasses, spyglasses—emphasizing that the same material can be used for diverse purposes, which is a good thing. Scientists and engineers are constantly devising new materials and new ways to make the chemical elements and their compounds into new and useful things, but there are still limits to what can be done at the atomic scale, as workers in the fields of microelectronics, micro-mechanical devices, and nanotechnology know so well.

Even though marvelous new things can be made out of the same old hundred-odd elements, that is not to say that they will be manufactured and become available at the corner store. Doing anything costs money, and doing exotic new things usually costs extraordinary amounts of money. If, for example, an ever-clean drinking glass could be made out of a clear Teflon-like material that shed germs like water, it might cost so much to manufacture that none but the very rich would be able to buy it. The cost of things places as much of a limitation on their availability as any chemical or physical factor. If even a near-perfect drinking glass could be produced, its price would be far from perfect.

Still, we can imagine such things. And though there may be no limits to what can be imagined, there are many practical limits to what can, in fact, be designed and made. To imagine the impossible, such as an absolutely perfect glass or a perpetual-motion machine, does not make it possible. But, conversely, the possible can never be realized if it is not first imagined. Yet, even though imagination may be more abundant than all the grains of sand and amounts of money in the world, that is not to say that the right idea will occur in the right mind in the right place at the right time. Even if we can imagine that the idea of a weapon of mass destruction might conceivably have occurred to Democritus, ancient Greek technology could not have sustained a Manhattan Project and thus the creation of a successful atom bomb.

As the examples and case studies in this book demonstrate from many points of view, design must always be done within a context, which makes some things difficult and others impossible—regardless of the time in history. A perpetual-motion machine is an impossibility because the behavior of all machines is constrained by the laws of physics, which simply do not allow for unchallenged motion. Even objects that are possible to make cannot be designed without consideration of the constraints imposed by their intended use. No matter what its design, a glass must hold water and stay upright on a stationary horizontal surface. Also, on being raised and tipped, it must release its contents into the mouth. A glass that will not do those things does not deserve to be called a drinking glass. A glass cannot be a theory that does not hold water.

Design also always involves choice, and choice limits use. A real glass must have a specific capacity. The designer must thus pick a specific size, or sizes, for a new line of glasses if it is to be more than a sketch on a tablecloth or an image on a computer screen. If the capacity of a glass is ten ounces, the glass can, of course, be used to hold any amount less than ten ounces, but never more. To hold more than ten ounces simply requires a glass of greater capacity. The designer, manufacturer, retailer, and consumer must all make choices about what size glass they will specify, make, stock, and buy. Such decisions will often be constrained by habit, custom, tradition, and practical considerations. A too-small glass has little use outside a dollhouse; a too-large glass ceases to be a glass at all: It becomes a vase, a bowl, or an umbrella stand.

Someone may wish to have a twelve-ounce glass because he has learned from experience that this is about how much water he will drink at one sitting at his desk. He has experienced having to refill a ten-ounce glass at inopportune times, and he has found that he never finishes all the water in a sixteen-ounce glass. Still, he has designed his desk experience around the ten-ounce glass because he once brushed his hand against the top of a larger one, spilling its contents over an important pile of papers. Since his cupboard had only seven-, ten-, and sixteen-ounce glasses, he looked for a twelve-ounce one in the housewares department at a local store. The salesperson told him that the store did not stock that size glass in the style he wanted and, furthermore, that the manufacturer's catalog did not even show it as something that could be special-ordered. The salesperson suggested another attractive style of glass, which did come in a twelve-ounce size. But that meant that the glasses in the cupboard at home would be mismatched. Since the new-style glass came in six-, ten-, twelve-, and sixteen-ounce sizes, a full set of new matched glasses could be used to replace the old, but then only a smaller number of each size could be accommodated in the cupboard. No matter what the decision, it involved a compromise. A matched set could not be maintained if only new twelve-ounce glasses were added, and the old glasses could not remain in the cupboard if a fully matched set of new ones was purchased. The most likely compromise seemed to be to continue using the old set, choosing between a ten- and a sixteen-ounce glass, depending upon whether the risk of a spill or the interruption for refilling was deemed more important or disruptive.

Design and its implications for daily life are ubiquitous. What style and size of glass we can buy, keep, and use depends upon design decisions made by any one of a variety of people. The conceptual design of the style of glassware must be done by someone—who may or may not have an artistic bent. A celebrity might be said to be the designer of a line of signature glassware for a housewares company, but, in fact, he may only have chosen one out of several designs presented by a consultant. An industrial or product designer working for a firm might be asked to come up with a new line of drinking glasses for a manufacturer, but the line would likely bear a name suggestive of the style rather than the designer. An architect might create "designer" glassware for a large retail chain and lend the status of his name to the line, and hence to the store. Whatever the provenance or name, consumers can only choose among those designs that become available, unless a designer is commissioned to produce something unique, a choice that could likely be made only at considerable cost.

Consumers are also designers. They make design decisions when they buy a house and when they choose what kinds of cars to keep in the garage. They make design decisions when picking patterns of china, silverware, stemware, and glassware for their house; when choosing furniture and furnishings; when selecting the quality, color, and monogram style for bathroom towels and bedroom linens; and when picking out artwork for the walls and plants for the hall. They make further design decisions when they purchase the clothes they wear and choose the restaurants they frequent in those clothes, and when they stay home to watch the television programs that they select or the movies that they have rented. They make design decisions about the brands of beer in the refrigerator and the types of wine in the rack and the kinds of snacks in the cupboard. They make design decisions every day and every night. Choosing a certain lifestyle is also a design decision, albeit one that can be tightly constrained by economic circumstances, thus limiting choices.

Because everyone makes so many unheralded design decisions every day, "nondesigners" are much more familiar with the nature of design than is commonly acknowledged. We all understand viscerally that design must always conform to constraint, must always require choice, and thus must always involve compromise. Understanding these elementary facts, we also understand why no design is perfect.

Knowing the imperfections of an object does not diminish it but, rather, can elevate the appreciation of the creative achievement that has minimized the intrusion of flaws into the design. Indeed, the bubble in the base that so dominates the design of my drinking glass distracts my eye from the irregularities of the glass's bottom. The last time my glass was emptied, I found myself looking alternately at the scene captured in the bubble and the full-scale scene beyond it. Doing so, I saw the glass itself in a different light. Now, when I look through the bottom of the (empty) glass, I can see that its surface is irregularly rippled and rough and contains a number of tiny air bubbles in the base. But, dispersed around the dominant bubble as they are, the little ones are as invisible to the casual drinker as the outer planets were to astronomers without telescopes. Whether or not the designer of the drinking glass intended the sun bubble to distract the eye from any tiny planets that might be discovered in the rough ecliptic of the bottom, it does so very effectively. It is an admirable detail of the glass's design, one that deals with the difficulty of forming such a thick base without unintended inclusions or irregularities.

With imperfections of manufacture, however subtle, detected in its bottom, I am now looking even more closely at the glass that just yesterday I had so admired for its seeming perfection of form. With the glass full of water, its sides had appeared to be of uniform thickness. Now empty and dry, the glass more readily reveals its true dimensions, refracting as it does the text of the manuscript page on which it sits. Rather than seeing the text distorted in a uniform way through the sides of the glass, I see it bent like a current flowing around boulders in a river. Rotating the glass in place on its base, I find that I can change the degree of distortion that I see, eliminating it completely when the glass is in a certain position. In other orientations, the nature of the distortion changes to one in which strata of text are inclined like bands of color and texture in the rock cliffs beside a highway. With the glass in my hand, I feel no irregularity on the outside surface, but the thickness is clearly not uniform, as I can tell by running my thumb and index finger simultaneously up and down the inside and the outside of the glass, as if using a micrometer. The irregularity that I feel indicates that the thickness of the sides is quite variable from bottom to top. Furthermore, this variation is different at different locations around the glass, explaining why I get different effects as I rotate it in place. The sides of the glass are, in fact, a continuum of lenses. Once noticed, these consequences of the manufacturing process are magnified.

The closer I have looked at what first seemed to be an almost per-

The bubble in the bottom of a glass captures its surroundings.

fectly formed glass, the more I have come to discover that it has imperfections of form, just as it has imperfections of function. Admittedly, this glass is a manufacturer's second, the set having been bought in an outlet store in Freeport, Maine. (The striking design had distracted me from remembering this at first.) But the glasses looked pretty good to my wife and me when we bought them, for what at the time did not seem like seconds prices, and they have served us well over the years. They have been remarkably resistant to scratching, chipping, and breaking, and they have kept their freshly bought appearance. It is assuredly quibbling for me to dwell on recently discovered minute faults in objects in which I found no fault before, but it is quibbling in the service of design. By better understanding how a dominant design feature, like the prominent bubble in the base of a glass, can distract the eye from flaws and faults of even a simple common object, we can better appreciate the nature of design and imperfection generally. Still, we should not lose sight of the big picture by focusing too closely on the little one captured in the bubble or its satellites.

Absolute perfection in a drinking glass or in any designed object is impossible. Though our glasses are seconds, even ones that passed inspection and were sold at full price must themselves have had some flaws, albeit virtually imperceptible ones. Their imperfections are just more subtle and more difficult to pick up with the naked eye, even the trained inspector's eye. The goal of good design and manufacturing is not to achieve perfection but to minimize imperfection and render it acceptable, unimportant, negligible, unnoticeable. Designers can achieve this best by keeping in mind the constraints of the design problem on which they are working, by making design choices that enable at least the most important of the constraints to be satisfied, by accepting the compromises that must necessarily accompany choice, and by incorporating strong design features that distract us from minor flaws. The resulting (imperfect) design is most commonly judged by how well it appears to be perfect in displaying its form and fulfilling its function at a reasonable price. Indeed, it is the irritant in the oyster that produces the pearl.

Robert Frost once said that he would "as soon write free verse as play tennis with the net down." Designing without constraint would be like playing baseball without fences or foul lines. Designing without choice would be like negotiating a maze with no alternative routes. Designing without compromise would be like having your cake and eating it, too. Designing without fault is impossible. The result would not be of this world.

The necessity of having to accept limitation and imperfection in things designed does not diminish them as admirable achievements or us as human beings. Imperfection is part of being human, and we can no more expect perfection of our inventions and designs than we can create perfection as inventors and designers. That is not to say that we should not strive for perfection, or that we must settle for gross imperfection. Good design, that which seeks a happy medium, will always make a half-empty glass appear to be half full.

THREE

Design, Design Everywhere

WATER WAS ONCE as free and democratic as rain. It filled clean streams and lakes, from which all people drank, scooping up mouthfuls in their cupped hands. Though most early camps and settlements would have been located near an adequate source of water, with the growth of towns and cities and their populations, it became increasingly necessary to bring more water to them. The Romans are famous for their aqueducts, some of which remain standing today as enduring symbols of civilization and as great monuments to ancient design and engineering. Indeed, the Pont du Gard and Segovia aqueducts attract bridge lovers and art historians alike among the throngs of visitors that marvel at the towering arches.

Getting water from abundant sources to where it did not occur naturally, or from where it was not needed to where it was, or even from hand to mouth, has always been an important problem. Its solution has affected the design of everything from barrels and bowls to tunnels and dams. Collecting the rain falling on roofs drove the development of the gutter, the drainpipe, and the cistern. Lifting water from the depths of the earth encouraged the development of wells and pulleys and pumps. Draining muddy water from mines was what drove the development of the steam engine.

Water did not have to look dirty to be dirty, however. Throughout the latter half of the nineteenth century, there was a growing awareness that water from all sources, including wells, was potentially contaminated with microorganisms that could lead to outbreaks of diseases like typhoid fever and cholera. Around the turn of the twentieth century, there arose a heightened sense of urgency to purify public water supplies. After Philadelphia put filters into use in 1906, the incidence of typhoid dropped by a factor of ten, and in another decade or so the disease was virtually eliminated due to further improvements in sanitary engineering.

No matter how sufficiently treated the water was, however, there remained another source of health risk from the use of common water supplies. Into the early twentieth century, the typical public water barrel, well, pump, or spigot had a communal cup, known as a sipper or dipper, from which everyone, the healthy and the sick alike, drank. The common sipper was seldom washed, and so it was the source of germs and disease. Because of an increasing awareness of the dangers of sharing a drinking utensil, the idea of providing sanitary drinking cups began to gain currency. But since it would have been impractical to expect users to wash a publicly shared drinking cup after each use, alternatives were sought.

The idea of making clean and disposable paper cups available at public water fountains gained currency at the beginning of the twentieth century, but not all who pursued the idea had purely altruistic motives. One group of Boston investors saw money to be made in such an enterprise, and they were looking to form a company to manufacture a drinking cup that could be dispensed from a vending machine in exchange for a coin. The investor group was planning to employ a design for a paper cup that would be stored and dispensed in a flat, folded configuration, and that could be opened to receive water from a nearby cooler. The idea made sense. After all, paper bags were stored in a flat, folded configuration and were opened only when they were used. Machines that made paper bags had long been operating, and it would have been no great leap of technology to make similar machines for the manufacture of folded paper cups.

Among those introduced to the investment scheme was Lawrence

W. Luellen, who in 1907 found out about it from a business neighbor who happened to be the lawyer representing the group of investors. Luellen was intrigued, though he thought the chances of success for the venture would be greatly enhanced if the cup were delivered round and fully formed, so that the customer did not have to unfold it to get a drink of water. This would likely mean a bulkier and more complicated dispensing machine, of course, but the trade-offs seemed worth it. Luellen, like many a designer, thus faced the problem of how to present to the investors the idea that was in his mind. He had to become an inventor, a developer, an engineer, an entrepreneur himself.

Luellen knew there were many ways to form a flat piece of paper into an open drinking cup. The folded-bag concept, derivative of the paper bag used for merchandise, was only one. His mind, however, went in a different direction. He envisioned starting with a single circular piece of paper and forming it into a round cup with pleated sides. His cup would not have been unlike the little cups that we use when sitting in a dentist's chair or even the smaller ones into which catsup and tartar sauce are dispensed at a seafood shack. Since paper absorbs water, Luellen thought of treating it with a waxy substance like paraffin, which would also serve to hold the pleats of the cup in place and give it some rigidity.

Another design that Luellen began working on was for a two-piece cup. In a patent entitled simply "Cup," it is described as being made preferably out of "water-proof paper" with a "tapering side wall . . . and a flat bottom portion." Not unlike the way a tinsmith might begin to make a metal cup, Luellen began with a flat piece of paper shaped somewhat like a contemporary shirt collar and curled it into a "frusto-conical form." This surface of a truncated cone constituted the sides of the cup, and a separate piece of paper attached to the smaller end of the frustum provided a flat bottom. The cup so formed did not have the rigidity of a pleated cup, but in time Luellen came up with the idea of forming a flange around the top edge, which served to stiffen the whole cup. In addition, this flange proved to be a handy detail to be engaged by a "suitable mechanism" within a dispensing machine so that one cup

Early paper cups were stiffened by a flat-flanged rim.

at a time could be disengaged from a stack of nested cups, another of Luellen's ideas. In 1915, the incorporation of a raised bottom helped to maintain a looser fit and an increased uniform spacing between the flanges of adjacent nested cups, thus providing for more reliable dispensing.

Along with developing his cup design, Luellen worked on designing the necessary infrastructure on which it would rely and within which it would function. He patented a manual "Cup Dispensing Device," whose central idea was to hold a stack of cups in place until the bottom one was pulled free for use. The flexibility of the paper cup was critical to proper operation, and today's paper-cup dispenser works on essentially the same principles embodied in Luellen's patent: When the bottom cup in a stack is pulled down, its rim deforms slightly to allow it to be tugged past pinch points on the inside surface of the dispenser. Those same pinch points serve to hold back the remaining cups, thus preventing them from falling out of the dispenser under the action of gravity alone. (Users had to learn to grasp the cup firmly but lightly, lest it be crushed in the process.) Not all paper cups were to be dispensed freely by just the grasp of a gentle hand, however, and so Luellen also patented a "Dispensing Apparatus," which required a coin to operate. Such dispensing machines were once widely found in train stations, theater lobbies, and other public places.

For the machinery to manufacture the paper cups themselves, Luellen sought the help of an engineer and inventor, Eugene H. Taylor, whose Taylor Machine Works was located in Hyde Park, Massachusetts. Taylor did design a machine to make the two-piece cups, but devising one to make pleated cups proved to be more difficult and costly, and so after a while that effort was curtailed.

Early in 1908, Luellen had come up with the idea for a combination watercooler and dispensing machine, which he called the Luellen Cup and Water Vendor. To later generations, it might have looked somewhat like the porcelain tower that stood beside a dentist's chair, but this "ancestor of the modern hot and cold drink vending machine" functioned not unlike today's stand-alone office watercooler. In its time, it was advertised as a vendor that "delivers to each and every person a new, clean cup filled with pure cold water for ONE CENT."

The American Water Supply Company of New England, formed to manufacture, install, and service the Luellen Vendor, was incorporated in Massachusetts in 1908. Soon seeking to expand the business, Luellen teamed up with his brother-in-law, Hugh Moore, who dropped out of Harvard and gave up plans of becoming a journalist in favor of this new and different kind of paper business. It was likely from Moore's typewriter that the purple text of the American Water Supply Company's brochure, *Quaff Nature's Nectar from* This *Chalice,* flowed.

The climate was just right for introducing the paper cup and singing its praises. At about the same time as Luellen and Moore were seeking to place their mechanical vendors in public places, the Anti-

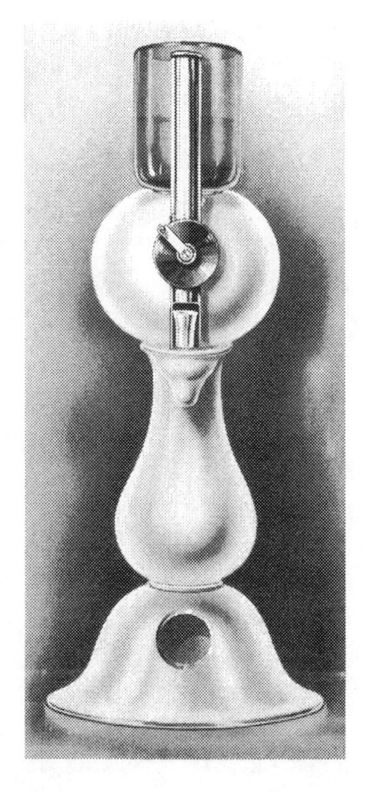

The Luellen Cup and Water Vendor provided cool, clean water in a sanitary paper cup.

Saloon League of New York had begun running advertisements promoting water machines as an alternative to saloons. According to the Reverend Howard H. Russell, founder of the American Anti-Saloon League:

> Thousands of persons are tempted into saloons every day because there are no sanitary drinking fountains where a man can get a drink of pure cold water in a clean cup. . . .
>
> It is a very sad fact that we have over 10,000 rum-shops in Greater New York, and not a single sanitary drinking fountain to be found. What we need in all cities is sanitary drinking fountains.

> . . . Each 1,000 *Luellen Cup and Water Vendors,* selling only 200 drinks each vendor per day, would mean 200,000 refreshing drinks per day instead of 200,000 temptations to drink something harmful.

Luellen and Moore no doubt made a similar calculation. At a penny a cup, they saw the same number of machines taking in two thousand dollars a day, or almost three quarters of a million dollars a year. Though neither the Anti-Saloon League's ad campaign nor anything else may have convinced many men to spend money for a cup of water rather than a glass of beer, other contemporaneous developments gave the paper cup plenty of impetus. In 1908, Dr. Alvin Davison, a biology professor at Lafayette College, carried out a study of the nature of contamination on common drinking cups used in the public schools of Easton, Pennsylvania. His article, entitled "Death in School Drinking Cups," described his experiments with one common cup, "which had been in use nine days in a school." Davison described his research method for examining this "clear, thin glass," and reported his findings about it in detail:

> It was broken into a number of pieces and properly stained for examination with a microscope magnifying 1,000 diameters. The human cells scraped from the lips of the drinkers were so numerous on the upper third of the glass that the head of a pin could not be placed anywhere without touching several of these bits of skin. The saliva, by running down on the inside of the glass had carried cells and bacteria to the bottom. . . .
>
> By counting the cells present on fifty different areas on the glass, as seen under the microscope, it was estimated that the cup contained over 20,000 human cells or bits of skin. As many as 150 germs were seen clinging to a single cell, and very few cells showed less than 10 germs. Between the cells were thousands of germs left there by the smears of saliva, deposited by the drinkers. Not less than one hundred thousand bacteria were present on every square inch of the glass.

Dr. Davison's article was soon being distributed by the Massachusetts State Board of Health. In Kansas, meanwhile, Dr. Samuel J. Crumbine, a public-health officer, argued for laws banning communal cups. State after state began passing legislation outlawing common drinking cups in public places, and a paper-cup company was well positioned to reap the benefits.

In 1909, Luellen and Moore formed the New York–based Public Cup Vendor Company, which focused on leasing machines for use on railroad trains and in railroad stations. The Lackawanna Railroad became the first to have the company install cup dispensers beside watercoolers on its trains, and both the railroad and the paper-cup company were eager to advertise the innovation. The Lackawanna had already established an advertising campaign that emphasized the use of the relatively clean-burning rock-hard "stone coal" known as anthracite in its locomotives. The Anthracite Line's spokeswoman was one Phoebe Snow, who by design dressed all in white and promoted the cleanliness and comfort of travel on the Lackawanna in ads that conveyed their message in rhymes that ended with variations on the phrase "upon the road of Anthracite." When the contract between the railroad and the Public Cup Vendor Company was finalized, Hugh Moore proposed a verse to sell the paper cups on the trains:

Phoebe dear you need not fear
To drink from cups that you find here
With cups of white no bugs will bite
Upon the road of Anthracite.

Moore's imagery was used in a modified stanza form in a Lackawanna ad announcing "individual drinking cups on all through trains":

On railroad trips
No other lips
Have touched the cup
That Phoebe sips.
Each cup of white

Makes drinking quite
A treat on Road
Of Anthracite.

Naturally, disposable paper cups were to vending and dispensing machines what disposable razor blades were to razors. Luellen and Moore, wittingly or not, had essentially adopted the marketing model that King Gillette had introduced only five years earlier. But the name Public Cup Vendor Company suggested that it was the dreaded public cups that were being vended, and so the name was soon changed to the Individual Drinking Cup Company, and the principal product named the Health Kup.

A flu epidemic following World War I gave the sale of paper cups a further boost, but with success had come increasing competition from other designs. One competitor of the Health Kup was the cone-shaped cup formed out of a single piece of paper, thus compromising shape for efficiency of manufacture. This familiar disposable drinking vessel was made by the aptly and snappily named Vortex Cup Company, which had been founded in 1912 in Chicago.

To compete with Vortex and other cup companies in the wake of the war, the Individual Drinking Cup Company looked for a similarly catchy name for its product. It so happened that next door to its factory was the Dixie Doll Company. Moore liked the look and sound of the word *Dixie,* and he recalled that in New Orleans it had referred to ten-dollar banknotes that bore the French word *dix* on their reverse. (The notes were issued before the Civil War by the Citizens' Bank and were said to come from "the land of Dixie." This designation, which at first referred mainly to New Orleans, soon came to refer to all of Louisiana and, eventually, to the entire South. The 1859 song "Dixie's Land" popularized the designation in the familiar forms of "Dixieland" and, finally, simply "Dixie.") In 1919, Moore renamed his Yankee company's product the Dixie cup. The Individual Drinking Cup Company and Vortex would eventually merge, in 1936, becoming the Dixie-Vortex Company, but in 1943 that name was shortened simply to the Dixie Cup Company, which was acquired by the American Can Company in 1957. By the beginning of the twenty-first century, Dixie was a division of the Fort James Corporation, a manufacturer of paper products.

In the early 1920s, ice-cream makers, who sold their product in bulk, were trying to reach a larger market—and so was the maker of Dixie cups. People could already buy soda and candy in single portions, so why not also sell ice cream in individual packages? Wax-coated paper cups, which for obvious reasons were not at all effective for holding hot drinks like coffee or tea, were the ideal new container for a cold product like ice cream. In time, a Dixie cup was developed with an all-around groove near the top of its inner circumference, into which fit a flat lid with a tab that could be pulled up to expose the ice cream within.

The ice-cream cup, like the paper drinking cup, might be said to have been "perfected," but that could easily have been argued by someone without a spoon to eat the ice cream. Like virtually all designs, the Dixie cup was not entire unto itself. With every Dixie cup—as the individual portion of ice cream itself soon came to be known—the customer had to be provided with something that served as a spoon. This adjunct to the Dixie cup came to take the form of a flat wooden device that was typically more bone- or paddle-shaped than spoon-shaped. To maintain the sanitary image of its origins, this typical spoon that I knew as a child was enclosed in an individual paper wrapper and detached from a ribbon of identically wrapped spoons when it was dispensed to the customer.

More recently, individual portions of yogurt have become popular as a quick and healthy breakfast or lunch. The container they come in is typically of plastic, as is the spoon that usually has to be supplied separately. One brand of yogurt has advanced the state of the container system by incorporating into its lid a plastic spoon of clever design. The halves of the spoon, which is made in two parts to fit into the three-inch-diameter lid, can be punched out from the lid and then fitted together to make an eating utensil that is just long enough to reach the bottom of the four-inch-deep container. Though the packaging system might now be said to be "perfected," it is not, of course. The spoon has to be assembled and can unsnap if used too vigorously. Furthermore, like virtually all spoons, which are rounded at the tip so as not to poke the sides of the eater's mouth, it is not very efficient for getting the last drop of yogurt out of the bottom of the container.

A bottle of water can be more easily emptied of its contents. That is not to say that it is as freely available as water from a public sipper or even from a penny Dixie drinking cup. Indeed, gallon for gallon, bottled water is generally more expensive than gasoline, but there is general agreement that it tastes better. It also tastes better than much, if not most, tap water, and so increasingly American restaurants have given diners the choice of local or bottled water, with or without carbonation. Many restaurants like the fashion of diners drinking bottled water, of course, because its cost, plus a handsome markup, can be added to the bill. (The habit of drinking bottled water has become so pervasive that even in cities known for their excellent tap water, like New York City, whose water is still brought down from upstate through the historic Croton aqueduct, diners reject the free stuff and order designer water.)

Drinking bottled water in America seems first to have become popular in the 1970s, when there was an increased awareness of the pollution of lakes and rivers and the contamination of water supplies. The craze for springwater took hold especially with the wide distribution and growing taste for the sparkling and fashionably French Perrier. Now, water bottled at many sources is sold everywhere from truck stops to museum shops. Just as sophisticated diners have their favorite wines, so now do most people have their favorite label of designer water, ordering it by name.

Plastic bottles of water also appear conspicuously on otherwise-formal tables and lecterns, captured in news photos of world summit conferences and on television broadcasts of major meetings and lectures. When I was young and played soldier in Brooklyn's Prospect Park, my friends and I would carry a supply of water with us in a metal canteen bought in an army surplus store. When our canteen was empty, we would refill it at a public water fountain, being careful first to use a dirty hand to rub any germs off the nozzle. (Some kids thought that you were supposed to suck the water out of the water fountain, a belief perhaps reinforced by the frequent occurrence of low water pressure in the park.)

Plastic bottles of water are now carried everywhere, and even the most sophisticated of ladies can be seen drinking out of bottles in a manner that only decades ago was associated with punks and public drunks. Such bottles may not be shared with others, but they must certainly collect their owner's own saliva and bacteria as they are drunk from and refilled, sometimes from public fountains, over the course of days. One can only wonder what Dr. Davison might have found if he had had the pieces of a plastic bottle to put under his microscope.

Water drinkers of all ages now eschew public fountains, which stand idle in many a school hallway as students and teachers alike have become accustomed to carrying their own plastic bottles. Knapsacks even have special pockets for the omnipresent water bottle. In fact, students and others generally have become so used to drinking water from a bottle that the public fountain may be in danger of becoming obsolete.

I can remember years ago having to stand in a queue to get a drink of water from fountains everywhere. Older school buildings often had porcelain fixtures attached to the wall, some bordered by elaborate tilework. In buildings constructed around the middle of the twentieth century, there were specially designed recesses in hallway walls, into which a square-bottomed, self-standing, chrome-encased water fountain was placed. The alcoves were outfitted with plumbing and electrical connections, so that the water could be cooled. The building in which I now have my office had such recesses, which kept the hallway clear when no one was drinking and which encouraged whatever lines of people that did form to keep close to the wall.

A few years ago, these perfectly fine water fountains were removed. Instead of being widened to accommodate wheelchairs, the recesses were plastered and painted over. New, baseless fountains have been installed on the flush wall, but these jut well out into the hall, obstructing the right-of-way. An electrical cord droops down from the underside of the fountain and plugs into the wall in a very unsightly way. The new fountains are wheelchair-accessible, but I have yet to see one chair-bound person use them. Indeed, few students—or anyone at all, for that matter—use the fountains now because so many people have become accustomed to carrying their bottled water with them. However, the fountains do create a bottleneck because of their placement. They thus provide an example of a design "improved" to the point of being inferior to the old. Though they do allow for handicapped users, the fountains are a hazard for people, on foot or in wheelchairs, who happen to be walking or wheeling close to the wall. Should the person be looking at and talking to a friend, he or she could easily run into the obstacle.

The popularity of bottled water has naturally extended to its use in the home. Supermarkets now devote a considerable amount of shelf space to a variety of sizes, types, and brands, but a week's supply tends to be bulky to transport from the supermarket to one's home and takes up a lot of room in the pantry. As an alternative, entrepreneurial suppliers will gladly provide regular deliveries of large refill bottles for home watercoolers. Ironically, a modern house connected by plumbing to a reliable source of fresh, safe, and abundant water can now devote about as much space to storing bottled water as a nineteenth-century farmhouse did to maintaining a rain barrel.

Because some, if not many, home water supplies still do leave something to be desired as far as taste, clarity, or hardness is concerned, a variety of water-treatment devices have come to be invented, designed, and developed. When my family lived near Chicago, our local water supply was considered by most residents to be too hard to use untreated for baths, showers, dishwashers, and washing machines. Common soaps and detergents were simply not very effective in the mineral-rich water. So most houses were fitted with water softeners, which were bulky devices that required regular attention to replenish the blocks of salt that their chemistry relied upon for operation. The one-cubic-foot salt blocks took up a lot of floor space near the supermarket checkout counter and were heavy to lug home and to drop into the water softener. Yet they did make the water more latherable and palatable.

Sometimes tap water, whether from a city line or a well, contains chemicals and minerals that are not necessarily dangerous but that impart a somewhat distasteful flavor. In this case, filtering can be effective in producing water that tastes as good as the bottled kind. One such device is the Brita pitcher, most models of which are smaller and lighter than a gallon jug and can be repeatedly refilled with tap water. Most styles of this pitcher have a white plastic top compartment, into which the water is run. It then flows slowly, sometimes agonizingly so, depending upon one's state of thirst, through a special replaceable filter (good, according to the manufacturer, for processing about forty gallons) and into the pitcher's clear plastic bottom, from which clean and good-tasting water can be dispensed.

The first Brita pitcher that my wife and I acquired did wonders for our summer home's well water, which when first used early in the season could be cloudy, smelly, and poor-tasting. The design of the pitcher was such that it had a removable plastic lid, which was prone to fall off if the pitcher was tipped over to get the last drops of water out. We became used to this after a while and made sure to keep a thumb on the end of the lid or hold it on with our other hand, as one would with the top of a teapot, while we poured. But the pitcher lid also had another annoying design feature. Because it needed to be removed completely during refilling, the cover either had to be put down on the kitchen counter or held in the hand not grasping the pitcher. Since the latter alternative left no hand free to turn the water tap on and off, the removable lid was a feature of the Brita that cried out for redesign.

Our second Brita pitcher, which we bought more recently, does have a redesigned top—one that snaps on and does not have to be removed completely during filling. The top has a sort of trapdoor in it that can easily be flipped up by depressing with one's thumb a lever on the pitcher's handle. It works easily and smoothly, and the pitcher requires only a single hand to hold and operate it during filling, leaving the other hand free to operate the faucet.

There is another, more subtle design flaw in the old-style Brita, however. I was reminded of it during the call-in radio show on the design of everyday objects, when a listener described her frustration in sharing the use of her Brita with her husband. If one of them tried to pour a glass of water just after the spouse had topped off the filling

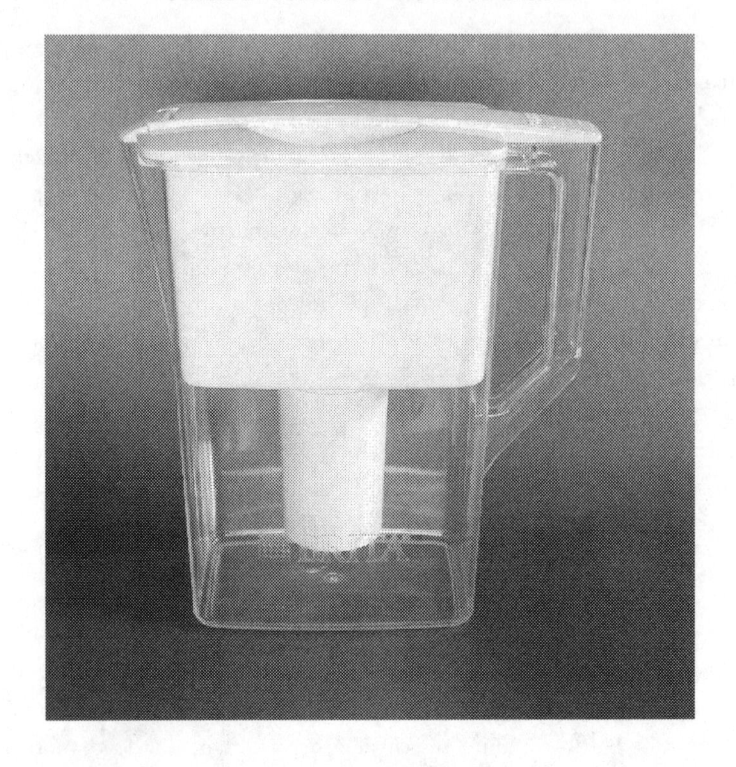

Older-style Brita pitchers had tops that needed to be removed for refilling.

compartment and put the pitcher back in the refrigerator, he or she would unwittingly pour a mixture of filtered and unfiltered water into the glass. The caller and most Brita users assume that the opaque top compartment is empty when they pick up the pitcher, and so if it has just been filled without their knowledge, they are taken by surprise. (They could be alerted to such a situation by the heaviness of the pitcher compared to its visible water level, but few people pay such close attention to the things they use several times a day.)

The annoying design feature could easily have been remedied by making the top compartment of the pitcher of clear plastic, like the bottom. Such a redesign would have ruined the symbolic asymmetry of the design, however. Presumably, the user would not want to see the dirty tap water in the top compartment of the pitcher. Furthermore, if the top compartment had been made clear like the bottom reservoir, the psychology of the design might not have been so effective. (A competing water-filtering pitcher does have a transparent top, but it is tinted blue, thus masking the color and clarity of the water.) Though some tap water may indeed be quite cloudy and dirty compared to filtered water, in many cases the visual difference is practically nil. If the filtered water did not look as if it was significantly clearer, would the pitcher system appear to be worth the investment in money and time? The filter, especially when new, is also prone to release carbon dust, some of which can pass into the bottom chamber but much of which actually floats up into the unfiltered water, making it look even more contaminated than it actually is. This is normal for the system, the Brita literature tells users, and the customer is advised not to use the first two pitchers of water that flow through a freshly installed filter: "Discard this water, or use it to water plants." The opaque top of the pitcher minimizes the effect of this aesthetic defect. As for a quick fix to the problem of sharing an old-style Brita water pitcher with a housemate, one of the participants on the radio show proposed living alone, a radical design change if ever there was one. The newer-style Brita eliminates the problem through the less drastic measure of having a more enclosed top, which does not allow the water to flow out so readily—unless the chamber is nearly full.

The design of anything is always difficult. There can simply be too many things to think about in even the simplest of devices and systems. Designers know that they should consider all of these things, but invariably some aspects of the design become the focus of attention and the others are overlooked, forgotten, or rationalized as being unimportant. We can imagine how the original Brita might have been designed. The removable top might not even have been a part of the prototypes tested in the laboratory. The top is largely a cosmetic lid over the guts of the pitcher, which is what the functional designers would likely have focused their attention upon. They would, no doubt, have been interested mostly in how large to make the filling compartment relative to the reservoir, how quickly the water passed through the filter, how effectively the filtering system worked, and other such functional matters.

Deciding on the final shape of the pitcher, including its top, would probably have been the task of industrial designers. Their interests would naturally have been focused more on how it might fit into the refrigerator, as well as on the aesthetic and psychological aspects of the new product. They would likely have paid more attention to color, texture, weight, and balance than to whether the top would be on or off during filling. That is not to say that designers of any kind should not consider such details. But for decades, refrigerator pitchers typically had removable lids, and so the feature may not have cried out for attention. It was not what was new about the Brita, after all.

The problem of the pitcher's filling and use by different members of a household in close succession would be a difficult one to catch in the design process. Designers are much more likely to try to imagine how a new design will be employed by a typical user. After all, a water pitcher is not supposed to be like a parlor game, in which players take turns manipulating the game apparatus in competition with one another. If the designers of the Brita did not anticipate what users in time found to be less desirable aspects of its design and operation, they had no malicious intent to frustrate users or to make early models less desirable than later ones. The evolution of the Brita pitcher top is representative of the evolution of product designs generally. It is precisely because faults and flaws come to the fore over the course of real use in actual situations, rather than in design-laboratory settings, that virtually all products change in minor details over time. Different choices and compromises, informed by the feedback from users of the prior design, are made within the same constraints.

Consumers are sometimes frustrated by this conscientious and usually well-intentioned effort to improve designs. Those consumers who buy an early version of a product learn to live with its little flaws and idiosyncrasies. The owners of early Brita pitchers figured out how to place its top lid upside down beside the sink so it stayed clean while the pitcher was being filled. They placed a thumb over the lid while pouring so that it did not fall off. They became good citizens of the household and replenished the pitcher after using it. They waited until the water charge had passed through the filter before pouring from a fresh pitcher. These were not difficult things to learn, and once learned, they became a natural part of using the device. The minor flaws in design became part of the product's mechanical personality. Frequent users no more noticed the faults than they did their spouses' tics.

If people who became accustomed to using the old design suddenly find themselves having to deal with a "new, improved design," they may experience more frustration than pleasure. The familiar routine of taking the lid off to fill the pitcher is thwarted by the fixed top. Putting one's thumb on the lid will now flip up the new fill door when the pitcher is in use. But it does not take long to learn that new habits of operation are in order. Unless someone is a stubborn reactionary who refuses to accept anything different, before very long there will be an appreciation for the fact that the new pitcher is indeed a "new, improved design."

Of course, the Brita pitcher or any other product may never be truly perfected. The problems of multiple users or changing fashion or new aesthetics will always be lying in wait to frustrate the product's designer, manufacturer, and owner alike, preventing them from reaching a complete equilibrium with regard to design and taste and function. Who can say that it will not soon become unfashionable to have a white plastic pitcher in the kitchen? Who knows when it will become environmentally correct to drink water straight from the tap?

It is also because the next wave of questions is not totally predictable that it is so difficult, if not impossible, to reach a final—unchanging or unchangeable—design. Virtually every design will be used by people other than those who created it, will exist in a context of ever-changing fashion and fad and other designs, and may potentially be shared and so not put back the way it was found. Every design will forever hold surprises and disappointments, as well as pleasures.

FOUR

Illuminating Design

STICKS AND STONES may occur in nature, but not in the form of levers and fulcrums. Those were the products of the minds and hands of our ancestors. Such invented things were originally conceived and designed out of nothing more than what was available, and whenever we use them now, in however slightly new a way, we redesign them. If we pick up a stick and lay it over a stone to move a rock, we make a new machine. We engage in the process of design by choosing that stick and that stone to move that rock, and in placing that stick upon that stone and under that rock in that particular way. Design is always with us and it always will be. It is at the same time ubiquitous and unique. It is common, but it is also personal. It can be as easy as ordering from a one-item menu, or it can be as impossible as effecting perpetual motion. No matter, it always involves constraint, choice, and compromise to solve the problem at hand. And though problems are always at hand, their design solutions will never stay put.

As designs that were copied or adapted from nature increased in number, our more recent ancestors acquired an accumulation of made things and ideas from which to draw inspiration for inventing and designing still more artificial things. Instead of continuing to sit on rocks, in time our ancestors made more comfortable chairs with seats of wood and animal skins. Rather than continuing to rely only on sunlight and moonlight and campfires for illumination at night, they devised torches and candles, oil lamps and gas lamps, and, eventually, electric bulbs. Instead of just having to remember the changing positions of the moon and stars and planets, they conceived methods of counting and recording cycles and positions. In time, they created calendars. Rather than relying on oral tradition, they developed means of keeping a written record of poems and history and rules of law. Thanks to our more recent ancestral inventors, designers, and engineers, instead of our having to go to a theater to be entertained, radio and television bring entertainment of a sort to us.

Today, a television, a chair, and a lamp, and maybe also a magazine or a book, might be said to furnish a room. But what size bulb should be put in the lamp? If we are planning to watch television all evening, we might prefer a 40-watt bulb, to provide just enough background light to keep us from tripping when we make our way to the kitchen for a snack during commercials. If we wish to check the evening's program guide quickly to see what is coming on at nine o'clock, we might prefer a 100-watt bulb to illuminate the newspaper page. And if we are going to read instead of watch television all evening, we might wish to put a 150-watt bulb in the lamp socket. As is the case with virtually every design situation, there is not a single choice that satisfies equally the needs for every different use.

We could, of course, install in our lamp the largest bulb we anticipate ever needing and also install a rheostat to control the brightness. This would, however, involve some installation time and expense that we might not wish to incur. It is the kind of capital outlay we might want to put off until brighter financial, and more leisurely, times. But no design problem has only one solution, and the least expensive can often appear to be the best investment for the time.

A scheme involving a modest expenditure for equipment but not of labor would be to keep a selection of lightbulbs beside the lamp and screw in the one that comes closest to satisfying the needs of the moment. How extensive such a selection might be would depend upon many factors. For one, lightbulbs come in only certain wattages. Usually, we can readily find bulbs of 40, 60, 75, 100, or 150 watts in the store, but we would waste a lot of time looking for one of, say, 55 watts. Besides, who among us would be able to distinguish between the level of illumination provided by a 55-watt bulb and a 60-watt one? Even if we could obtain lightbulbs in increments of 2- or 3- or 5-watt ratings, could we say what would be the lowest wattage we would expect to need, and what the highest? Would we ever really need a 2-watt bulb, or a 750-watt one? Even if such a wide range were available to us, how many bulbs could we practically and safely keep beside a reading lamp, which might be rated for only so many watts anyway? And how would we find the one perfect bulb when we needed it? Not every scheme that can be thought up is possible or practical to realize. So we might compromise and just keep three bulbs beside our lamp: one each of 40, 100, and 150 watts.

But even with this selection, we would have a problem every time we wanted to change from one bulb to another. The change would necessitate removing the inappropriate bulb first, which would be inconvenient, cumbersome, and possibly dangerous. Unscrewing a bulb that has been on for some time often leads to burned fingers. Even if it can be unscrewed with some quick flicks of the wrist, the still-warm bulb usually has to be juggled like a hot potato to a resting place. It very well may crash-land, producing a mess of sharp edges, or at least a broken filament, and so the need for a replacement. Given this scenario, most of us might settle for a further compromise: to put a 100-watt bulb in the lamp and leave it, accepting the room being a bit too bright for watching television and a bit too dim for reading. But of such situations, inventions are born. Wouldn't it be nice, someone once must have thought, if the same lamp or lightbulb could provide different degrees of illumination at different times? Out of such a question might have come the design for a lamp with three sockets. If one socket holds a 40-watt bulb and two hold 60-watt bulbs, then we can have a choice of 40, 60, 100, 120, or 160 watts' worth of illumination by switching on one, two, or all three of the bulbs.

A more sophisticated and convenient solution is the lamp with a single bulb in a single socket that can be set to provide different amounts of illumination. The three-way bulb is a clever invention, a clever design. By incorporating distinct 50-watt and 100-watt filaments into a single bulb, three levels of illumination may be achieved, depending on whether one or the other or both of the filaments are connected through the electrical circuit. The three settings are often sufficient to illuminate the room at an appropriate level for most activities, as long as we are willing to accept one of the three discrete choices. This would not seem to be much of a sacrifice, for many people appear not to be able to distinguish easily even between two bulb settings 50 watts apart. This is borne out by the common practice when using a three-way bulb of returning the lamp switch to the dark position and then counting off the clicks of the switch to illuminate the bulb at the three distinct wattage levels. We seem to trust our ears more than our eyes in determining the brightness of a lamp.

The design problem of providing the proper level of illumination in a room becomes further complicated if the space is occupied by two people doing two different things at the same time. If one wishes to watch television with 50 watts of ambient lighting but the other wants to do needlepoint under 150 watts of lamplight, there is a clear conflict of desires. The obvious compromise of setting the light level at about 100 watts satisfies neither party completely. Depending on the dynamics of their relationship and the size of the house, the television watcher or the needlepointer may retreat to another room and another, more appropriately set lamp. The complexity of the design problem naturally increases with the number of people and the number of different things they are doing. A civilized and functional group will be able to compromise, as they implicitly do in public places, where they cannot control the lighting or much else. The dysfunctional group will demand concessions of one another they would not dare ask of strangers.

The world and its designed contents and activities work because we, its inhabitants, by and large accept constraint as we do the laws of nature, assess choice as we do value, and recognize compromise as we do civility. There is no sense in seeking unfettered choice, because choice is always constrained. There is no realizing perfection, because choice always requires compromise. The more we understand the nature of design, the more we understand why all designs are necessarily flawed, though not necessarily bad. It is just that there will always be room for improvement in whatever we make and do, whether in the privacy of our living room or in the openness of a convertible.

A convertible is a car with a light-switch roof: It is either up or down, on or off. Trying to drive down the highway at full speed with the roof in an intermediate position is to invite the wind to rip it off, making a two-way roof into a no-way one. When properly operated in the proper climate, the convertible car roof can bring a lot of pleasure to sun worshipers and windsurfers. However, when the weather is wet or cold, the roof must be raised, and in that position, it can be annoying to those who prefer the unfettered feel. Especially in older convertibles, the fabric of the soft top was susceptible to being torn or cut, making for a wet and cold experience. It also tended to flap or flutter in the wind, making for a noisy ride.

In 1955, Ford tried to improve on the convertible by designing a Fairlane Crown Victoria Skyliner model that came with a transparent plastic roof panel over the front seat. It offered "the liberating feel of top-down motoring without the hair-mussing inconvenience of a true convertible." The design has been described as "a noble, albeit flawed, attempt at the elusive ideal." Though Ford had tested the car in the desert and claimed that its interior heated up only about five degrees more than a conventional hardtop, customers complained that driving in it could "seem ovenlike." A snap-in vinyl sunscreen was offered, but the car did not live up to its promise.

In 1957, Ford tried to improve on the cloth-top convertible in another way. The Fairlane 500 Skyliner came with a retractable hard top, which required a shortened roof design that had to be hinged to fit into an elongated trunk. The trunk itself opened forward, and the spare tire had to be mounted behind the car, Continental-style. Not surprisingly, the car, which cost four hundred dollars more than Ford's conventional Sunliner convertible and had little space for luggage with the top packed away, proved "complicated and somewhat trouble prone."

The sunroof, when provision is made to keep the hot sun out, is a compromise between a fixed and convertible car roof. Although giving only a relatively small opening to the sky, the modern sunroof provides the option of letting a considerable amount of sunlight into the car while keeping the rain, wind, heat, cold, and noise out. Still, sunroofs can be opened on pleasant days to admit fresh air. The sunroof can, however, require a sacrifice of actual headroom in order to accommodate the sliding roof panel and opening mechanism, and this can make a difference for a tall driver. My wife and I were once assigned a rental car with a sunroof, and it was not easy for me to get into the driver's seat. At first, I assumed the seat was adjusted for a shorter person. We drove off, with me hunched over the steering wheel and fiddling around with the seat-adjustment controls while my wife worked at opening the sunroof. After driving a few blocks, I concluded that the seat was, in fact, as low as it would go, but my head was still pushed down by the car's ceiling. A crucial few inches of headroom had been sacrificed for the sunroof and its mechanism. We exchanged the car for one with a solid roof.

More important than letting natural light into a car's interior is projecting artificial light out front so that the vehicle can be driven in the dark. For many years, headlights were aimed by removing the chromed rims and turning a few mounting screws until the light beams hit the garage wall at the correct height and with the correct spacing between them. When in the car, the driver controlled two settings, those for low and high beams. The low-beam setting illuminated the roadway for only a short distance in front of the car, but in this position, the headlights did not glare into the eyes of the driver in an oncoming car. Dimming the lights was a social compromise between approaching drivers' visibility. Because a driver behind low beams could see only a short distance ahead, as soon as an oncoming car passed by, he pressed a foot switch to change to the high beams, which threw light far ahead on the highway. Driving on a quiet two-lane road was punctuated by the constant *click-click* of the headlight switch as cars approached and passed in the night.

Changing from high to low beams was an expected courtesy of the road. Without this model of social design, the oncoming driver would be blinded and blind another in turn. With dimming switches, strangers who would never actually see each other could communicate as if they were familiar acquaintances tipping their hats on a small-town sidewalk. When an approaching driver clicked his headlights down to the low-beam setting, the recipient of the gesture followed suit immediately, assuming that the glare of her car had bothered him even before his had bothered her. It was a rule of the road violated only at one's peril. A floor-mounted switch meant that the hands were free to hold the wheel or be ready to downshift if necessary. But because the switch was mounted on the floor, it meant that the left foot had to control both the headlights and the clutch. Luxury cars, like Cadillacs, which already came with automatic transmissions, began to come equipped with automatic headlight detectors, which switched the headlights to low beam as soon as approaching lights were sensed.

However, the glare of headlights remained a problem for cars driving in the same direction. Before day-night mirrors became standard, high beams from the car behind, reflected in the rearview and side mirrors, could result in a driver being blinded by traffic coming from both directions. When the driver ahead was being bothered, about all he could do was put his hand over the mirror or turn it to the side. The offense could be clearly seen by the driver behind, since her high beams were already illuminating the entire interior of the car ahead. The right thing to do was to lower the beams and rely on the car ahead to lead the way in the dark. It was also expected that low beams would continue to be used while overtaking and passing a car, the high beams being switched back on only when the cars were abreast of each other. Not every driver was that attentive or courteous, however.

When day-night rearview mirrors were introduced, they dealt only partially with annoying high beams coming from behind, failing to address the problem of high beams reflecting in the side-view mirror. Drivers who gave up on the social compact directed their side-view mirrors down to the ground, but this defensive adjustment made them useless as mirrors.

In time, at least in America, the automatic transmission became a popular option and then standard equipment in new cars, allowing the left foot to be dedicated to the high-beam switch. Now, of course, with that switch moved to the steering column, the left foot has nothing at all to do. Indeed, with the right hand freed from having to switch gears at differing speeds, the steering column became the location of choice for levers and knobs of all kinds, for everything from turn signals and windshield wipers to headlights and cruise control. With the many controls, automatic settings, and other gadgets available on newer automobiles, some drivers seem to have forgotten that in most cases they still are responsible for how their lights affect traffic coming and going.

Today, many drivers appear to have become insensitive to the fact that theirs is not the only car on the road. Increasingly, bright headlights seem to be locked in a high-beam setting. Drivers of sport-utility vehicles fitted with in-your-face high-intensity-discharge headlights, known, ironically, as HID lights, appear especially given to not hiding their high beams. Even when apparently properly aimed, these headlights can blind other drivers on the road, because aiming and driving conditions are different. The aiming of headlights is usually done when the vehicle is empty, but when it is fully loaded with passengers or luggage or purchases from the mall that lower the back of the vehicle, the headlights get directed upward and into the eyes and mirrors of drivers ahead. This has long been a problem, of course, since headlights could only be aimed for a certain loading of the car, but its effect is aggravated with the newer, brighter headlights and higher vehicles. The problem can be addressed by the introduction of automatic leveling devices, which, regardless of the car's attitude, will keep the headlights aimed at the proper inclination relative to the road. However, most older cars cannot be expected to be retrofitted with such devices, and so headlight glare will likely remain a problem on the American road for some years to come.

The situation appears to be different in Europe, however. On a recent driving trip through France, I was struck by the paucity of sport-utility vehicles on the road and thus by the absence of the attendant annoyance of having their headlights reflected in my rearview mirror. It was only upon returning to my office and reading through the accumulated magazines that I came to a realization: Apparently, I had not been so bothered by headlights in Europe because of a cultural difference in how we view technology and how such cultural differences affect designs. There are distinctly different regulatory standards prevailing in Europe and the United States when it comes to how headlights are aimed. Simply put, European regulators give a great deal of consideration to how the glare of headlights affects other drivers, whereas American regulators put the emphasis on maximizing the vision of the driver on whose car the headlights are mounted. This reflects a fundamental difference in European and American approaches to many aspects of law, regulation, and public policy. In France, which has a tradition of Napoleonic law, to be legal, something must be specified as permitted. Thus a headlight standard spells out what is allowable. If some new device is not stated to be allowable, it can be put on a car only at the owner's peril. If it is allowable, it can only be installed in a specified way. In the United States, on the other hand, regulations tend to be stated in terms of what must be minimum equipment on a car. The vehicle must have such things as emission controls, taillights at a certain height, and headlights that illuminate a given range on the road ahead. Anything that is not explicitly banned by American regulations can be put on a car.

The design of anything, from a lightbulb to how it illuminates a room, from a headlight to how it illuminates the road, involves an acknowledgment of limitations: Lightbulbs are made and sold only in certain wattages, and we must choose from among these the one, two, or three bulbs that will come closest to giving us a reasonable and acceptable level of light; conventional headlights provide us with only limited choices for illuminating the road. The use of such imperfect on-off technology in the context of society demands that we compromise and cooperate so that we can maximize benefits and minimize annoyances. There is no perfect headlight or lightbulb, and there is no perfect idea.

There are, however, claims to perfection. The Maglite, the sleek aluminum-cased flashlight invented by precision machinist Tony Maglica in his garage shop, seemed to be the ultimate handheld portable light source when it was introduced in 1979. The new device first caught the attention of police officers, who appreciated not only its superior illumination of crime scenes but also its hard casing, which was ideal for everything from bashing in windows to subduing criminals. Soon the Maglite was being bought by people in all walks of life as the flashlight for all seasons, and it became the industry standard. It threw an adjustable beam of bright light that was marred in only one detail—the so-called black hole that appeared in the center of the beam for most settings. Maglica believed that the black hole was impossible to eliminate completely: It was a physical fact that the source of illumination—the bulb—got in the way of its own light bouncing off the reflector. It was considered part of the makeup of flashlights of all kinds.

About a decade after their introduction, Maglites had their reputation blemished for another reason. Tony Maglica and Claire Maglica, the woman whom he had been living with for twenty years and who had taken his name, had a falling-out over who was going to inherit his business. He had planned to leave it to his natural children, whereas she expected it to go to her and her two sons, whom Maglica had raised as if they were his own. Indeed, since they were in high school, Christopher and Stephen Halasz had been made privy to every aspect of the flashlight business—Christopher rising to the position of director of research and development for Maglite, and Stephen becoming a vice president and later the company's outside legal counsel. However, after Claire Maglica filed a palimony suit and the trial was covered in all its nasty details on Court TV, her boys became personae non gratae at Maglite and so eventually struck out on their own.

In the late 1990s, the Halasz brothers "set their sights on creating the perfect flashlight," by eliminating what they saw as the Maglite's major flaw—the black hole in the light beam. They knew the source of the hole to be the familiar parabola-shaped reflector behind the bulb, and so they developed a deeper and more steeply sloped C-shaped reflector. To further distinguish their product from the Maglite, the brothers made their new flashlight's case out of plastic and marketed it to outdoorsmen under the name Bison Sportslight.

Tony Maglica saw the Bison as a direct competitor to his Maglite, however, and he sued the Halasz brothers for "trying to steal employees from his company and pilfering trade secrets they learned while working for him." He insisted that "the brothers' claim to have developed a superior flashlight cost him business," and he challenged as false advertising "their assertion that the Bison Sportslight emitted a beam with no black hole." The Maglite inventor asserted that he had "spent hundreds of thousands of dollars trying to get rid of the black spot, the black hole," but, he contended, it was impossible to do so. "You can get some improvement, but you can't change the laws of physics."

At one point in the trial, Maglica's lawyer clicked on a Bison Sportslight and adjusted the lens setting to demonstrate what the jury evidently saw as a black hole, for "they later found that the brothers' claim to have overcome the optical effect was false." The jury also "awarded Maglica $1.2 million to compensate him for lost business and to punish the Halasz brothers for stealing research information on bulbs, switches and electrical circuits." The brothers planned to appeal, asserting that the courtroom demonstration used "an extreme setting that no user would select" on their flashlight. That may be, but even the slightest imperfection can be hard to miss when put in the spotlight.

FIVE

Driven by Design

DESIGN COMES with a lot of baggage; it is the rare activity that has no impedimenta. Every design problem is laden with constraints and preexisting conditions and associations, which must be complied with and accommodated in some way. Every new thing becomes part of something else, and by its very introduction alters how the original thing is used.

The cup holder has become an important feature in the American automobile. Car designers and salesmen have come to recognize this, and the cup holder can be among the first details that someone buying a new car is shown. Indeed, such a focus on the cup holder may be a necessity in the showroom, for these days it is not unlikely that someone shopping for a car is carrying a bottle of water, a can of soda, or a cup of coffee. The refreshment needs to be set down somewhere if the customer is to be able to get the feel of the steering wheel and try the radio buttons in the car on display. Of course, it was not always this way.

In the early days of the automobile, food and drink were expected to be carried in a picnic hamper, and eating and drinking were done outside the car, usually on a blanket spread on terra firma. As roads were developed to accommodate increasing numbers of automobiles and tourists, roadside parks with picnic tables became increasingly common, further insinuating the importance of food and drink in the grand scheme of things automotive.

In time, drive-in restaurants and theaters appeared, and streamlined automobiles lined up in these areas like iron filings in a magnetic field. Eating and drinking in the car became popular, but it was done with the vehicle in a parked position. At the drive-in restaurant, a parked car was served by a carhop, sometimes on roller skates and almost always sporting a ponytail. She brought the order on a tray whose crooked arms hooked onto a partly open window and whose legs rested against the outside of the car, thus making the tray surface more or less level. Drive-in habitués knew exactly how far to roll the window up or down, and the driver was used to distributing the hamburgers and french fries to the passengers. Drinks were left on the tray until needed, or were nestled between the legs or set down on the floor.

At the typical drive-in theater, there were no carhops, and it was the movie's inside speaker, instead of an outside tray table, that was hung on the window. The experienced drive-in moviegoer rolled the window as far down as possible in the summer and as far up as possible in the winter. Popcorn and soda were sold at the concession stand, to which someone in the car walked during intermission or at a slow part in the film. The door of the glove compartment, if it opened flat, might occasionally have been used as a small tray table, on which those in the front seat could set their drinks. Since the small door would not have enough room for the drinks of the backseat occupants, too, they used the seat or floor and, at least until their drinks were finished, tried to remember not to shift around too much during the movie. Even in a car tethered to the speaker pole, excessive activity could bounce and rock the vehicle, causing the drinks to move to the edge of the glove compartment's door and topple over onto the floor. Manufacturers began to put shallow circular depressions in the back of glove compartment doors, not unlike those in the tray tables on airplanes. These recesses were primitive cup holders, but they were of little better use when driving over a rough road than an airplane's are during turbulent conditions.

The first widely available true cup holders were holsterlike plastic ones. These were hung onto a car door, but not over the top of the window glass. The cup holders had a thin, flat, hooked extension that was inserted between the window and the inside door panel, squeezed in between the glass and the then commonly used feltlike material that kept car windows from rattling. These holders came in the gaudy colors that so many plastic products still do, and they looked more than anything else like toys a child might play with in the sand at the beach. Nevertheless, with the wide distribution of such cup holders, which seem to have blossomed coincident with the introduction of pop-top aluminum beverage cans, open drinks could be carried in a moving vehicle with a reduced risk of spilling. (Special precautions had to be taken when opening and closing doors, however.) A driver who thought the cup holder clashed with his car's interior or interfered with window operation would have to hold an open can between his legs.

Commuters, who had grown accustomed to having something to drink in their cars, increasingly began to carry coffee with them in the morning. The problem with coffee is that it can be too hot to hold, much less drink immediately, and dangerous to spill. In the early days of space travel, some automobile coffee cups were made to resemble the shape of the Mercury capsules that carried early astronauts into orbit. The cup's wide base, padded with foam and designed to be set down on a broad, flat dashboard surface or the floor, kept it from tipping over. But an unattended cup in a car, no matter how space-age in design, is not able to withstand the g-forces that accompany sharp turns and rapid acceleration or deceleration.

With the proliferation of drive-through windows at fast-food places, Americans, at least, came to regard a place to park their cup as somewhat of an automotive necessity. Increasingly, car-bound customers were handed drinks without them being secured in a pressed-paper cup tray, which was basically a portable cup holder. The built-in drink holder was widely available in American cars by the mid-1990s, and it became customary for cars to come with multiple cup holders. When the 1997 Chevrolet Venture minivan was introduced, among its remarkable features were seventeen cup holders. Since the vehicle was designed to seat only eight people, each passenger could commandeer at least two drink receptacles. The cup holder in the 2001 Chrysler RS minivan was designed for versatility: Its design "allows it to expand or contract to accommodate 32 sizes of drink container—everything from a child's juice box to a Big Gulp. The assembly can even pivot to remain level when the seats are moved. . . ."

Since neither our 1980 Volvo nor our 1987 Ford Taurus had a cup holder, it was the furthest thing from our minds when my wife and I were shopping for a new car in the mid-1990s. Our family car became a 1996 Volvo, which we bought after sitting in many contemporary models. Since most of my time spent in the car is on long-distance trips, a principal criterion for me was that the driver's seat be comfortable. I sat in a lot of cars, rejecting most of them because they did not provide enough headroom, even with the seat at its lowest position. The Volvo did have the requisite headroom, though only barely, and my wife and I decided on it.

After having committed to a model (the 850 sedan), we were confronted with decisions about options packages, which came in odd combinations of features. To get a compass and a thermometer that told the outside temperature, we would also have had to take small wipers on the headlights. The former features seemed desirable and practical; the latter seemed frivolous. Not wanting to appear frivolous in what we considered an immensely practical car, we turned down this triple option. (Many a time since, especially when disoriented on rural roads, we have missed possessing a compass in our car.)

Now tired of looking for the perfect car and having come to the dealer that day committed to buying a Volvo, we ordered one with few options. We took possession a couple of weeks later and sat in the backseat of our new car while the salesman showed us what each of the buttons, knobs, and toggle switches did. It was a warm June day, and so he first concentrated on the controls for the air-conditioning, which worked well. One of the features of the car was a radio antenna that was extended only when the radio was on. Since the antenna was mounted on the left rear fender, it was out of sight of the driver. We discovered very shortly after bringing the car home that when up, the antenna just passed under our fully opened overhead garage door. We discovered this while driving into the garage as the automatic door was still moving into its fully opened position, and the antenna struck it. The slightly bent antenna proved to be an easy thing to straighten, but ever since the incident, we wait for the garage door to stop moving before we drive under it.

Some curious features of our new Volvo only became apparent to us after some weeks. Since the climate in North Carolina is typically hot and humid in the summer, we drove our new car with its windows closed. One day, while motoring on the highway, we had to pay a toll, and as we approached the tollbooth, I reached instinctively for the window button on the driver's door. Not finding it in the place I had become accustomed to on our Taurus, I moved my fingers up and down along the armrest without taking my eyes off the traffic jockeying for position ahead. Just a few cars away from the tollbooth, I looked down and did a double take when I saw no controls at all on the door. Disoriented, I was further startled when the window opened without my intervention. How did the car know when to open the window? I wondered. Actually, my wife had pushed a button on the console between us, and she closed the window after I paid the toll. Having all the controls between the driver and navigator has its advantages, but it also clearly has its drawbacks.

Since the car's console contains the handle for the automatic gearshift, it was something that I had looked at many times in getting used to the positions for drive and reverse. I must have seen the window and mirror controls time and again, but their location had not registered. Now, years after having bought the car, I still occasionally find myself reaching for the door first whenever I need to open the window or adjust the side mirrors. I usually catch myself quickly and immediately switch my grip on the steering wheel from my right to my left hand. With the Taurus, which I continue to drive back and forth to my office, I had become accustomed to holding the steering wheel with my right hand, thus leaving my left free to push the buttons to lower the windows or adjust the mirrors. At times, I thought that the Taurus must have been designed by or for a left-hander, but that foolish assumption was belied by the position of the gearshift handle and other controls. All automobiles require a good deal of ambidexterity to operate.

As with many a car these days, the console between the front seats of our Volvo extends back and up into an armrest, which also serves as the cover to a compartment for the little things that collect in cars. Depending on when a car was made, the console compartment may have slots for eight-track tapes, audiocassettes, compact discs, or videotapes. But whatever the car's vintage, the compartment is convenient, even if it draws things to it as a black hole does light. The glove and other compartments in our cars have always quickly become filled with necessities, near necessities, and nonnecessities. To get at any of them in the console compartment, we obviously have to remove our arms from its cushioned top, which is no big annoyance, especially since at least one of us often has to use that same arm to reach into the compartment for the desired object. It's a sensible design that gives a thing more than one function. In fact, in a Volvo the era of ours, the armrest/compartment cover has also a third purpose, and that is to store and support a pair of tandem cup holders, which we did not discover until trying to open the console compartment one day. Not being yet fully familiar with the location of its latch mechanism, I cupped my fingers under the right side instead of the left and, as if pulling a rabbit out of a hat, found myself deploying a pair of cup holders.

Since we did not carry drinks around when shopping for a car, and since we had not asked about cup holders, the salesman did not bother to tell us about them. (Only later did I come to realize that introducing cup holders into Volvos had been a controversial issue.) Now that we knew we had holders, we decided to use them. This was easier said than done. We actually had to learn how to use our cup holders, and only in time did we come to understand fully their capabilities and limitations. When not in use, our Volvo's cup holders are concealed within the cover of the storage compartment/armrest, like a hand up a sleeve. A pull on an inconspicuous handle extends the cup holders out in front of the armrest, certainly a convenient location for a drink. The black plastic accessory has two spring-loaded rings, each of whose diameter when fully deployed is slightly larger than a standard aluminum beverage can. Flexible tabs help stop cans from rattling in the holder. To keep cylindrical drink containers from dropping completely through the rings, there are spring-loaded plastic fingers that drop down and provide support from the bottom. The overall design of the cup holders is minimal: light, airy, and skeletal. It is also clever, and the space in the armrest where the device is contained is narrower and shallower than the holders' deployed size.

As neat a design as the Volvo cup-holder device is in concept, it proved to be terrible when put to actual use. When pulled out of the armrest, the cup holders sit directly over and not too far above the part of the console holding the window and mirror controls. When drinks are in place in the holder, there is barely room enough to reach down to adjust the side mirrors or open the window to pay a toll. Furthermore, with drinks being directly above these controls, there is a real danger of liquids spilling onto them and into the console. (One time on a long airline flight, a child behind us spilled a drink onto the control console between two seats, shorting out the power, and our entire side of the cabin was without light or fresh air for half of the flight.)

Cup holders deploy directly over the window and mirror controls in the 1996 Volvo model 850.

Once a design becomes the object of criticism, there seems to be no end to shortcomings that can be found in it. (The design critic becomes like a starved animal tearing at the flesh of its prey.) On closer scrutiny, the clever Volvo cup holders are not only poorly positioned but also poorly sized and configured. In order to allow what space there is for driver or passenger to reach under the cups to operate the console controls, the recess for the drinks is very shallow, barely a third of a standard can's height. As we have learned, there is a distinct hazard that a drink may be uplifted out of the recess or overturned by a hand reaching to adjust a mirror, open a window, or just moving to stretch on a long trip. Also, should the driver unthinkingly open the compartment cover to retrieve something, the attached cup-holder device lifts like a catapult arm and flings whatever is in it into the backseat. As long as these caveats are kept in mind, the device can function acceptably.

But *cup holder* is really a misnomer for a device that can accommodate only aluminum cans, small drink cups, and slender plastic water bottles. The larger cups and bottles ubiquitous in the convenience stores at many modern gas stations and rest stops beside interstate highways do not fit. Small paper and Styrofoam coffee cups do fit into the holder receptacles, but most larger ones will not. When two small coffee cups are in place in the closely spaced rings, the rims of the cups push against each other. This not only keeps them from remaining absolutely vertical but also makes it difficult to take up one without disturbing the other. In short, our Volvo's cup holders leave a lot to be desired. In a car that has so many thoughtful design details, like a left-foot rest in a car without a clutch, this was a mystery. Cup holders in automobiles were not part of traditional Swedish culture, however, and Volvo seems only reluctantly to have developed a retrofit for the American market. Whether the company liked it or not, increasingly in the 1990s, American car buyers were expecting something to hold their drinks. So something had to be designed to fit into the existing interior configuration of Volvo cars.

To appreciate fully the difficulties faced by the cup-holder design team, we can put ourselves in the driver's seat of an earlier Volvo model, one without a cup holder, and imagine trying to solve the problem ourselves. If there was to be a cup holder added, where could it be put? To the driver's left is the door. Attaching a cup holder to it would be impractical, because every time the door was opened or closed, the drink would slosh out of its container. The space on the dashboard is pretty much fully occupied by dials, gauges, warning lights, switches, and buttons. Few car buyers would be willing to give up a radio or CD player for a cup holder. (Interestingly, though, computer jargon for a CD-ROM drawer, and by extension a CD-ROM drive itself, is *cup holder,* a usage supposedly traceable to a call received on a help line from someone who wanted to know how to fix what he referred to as his computer's broken "cup holder.") In the mid-1990s, the only logical place to look to install a cup holder in a Volvo sold in America was in the vicinity of the console storage compartment.

Given that the console controls were in place, it was natural to look to secure the air rights above them. Though certainly astute designers could recognize that this would impede access to the controls and introduce the danger of spilling liquids over them, the acceptance of these disadvantages and risks could be rationalized for the sake of a cup holder for the gauche Americans. But the skeletal cup holder could not sit permanently over the controls, for that would present an eyesore and a clearly objectionable feature to the prospective car buyer. Besides, some car owners forbid drinks to be brought into their car and so would see a conspicuous cup holder as an unwanted invitation rather than a positive feature. Thus, the cup holder had to be there and yet not be there. Concealing the device in the console armrest presented the best of both worlds. For those who did not want any cup holder whatsoever, it could be kept out of sight and so out of mind, never acknowledged and certainly never deployed. To those who did want a cup holder, it was, like a sword in a scabbard, at the ready.

A decision to conceal the cup holder in the console cover would present the design team with some clear constraints. The height of the armrest could not be increased significantly, because then its padded top would be uncomfortably elevated for its primary intended use. The bottom of the cover into which the device had to fit could not be lowered appreciably, lest storage space in the already-small compartment be sacrificed and the deployed cup holder be so low over the controls that they could not be reached at all. The length of the console cover also had to be pretty much fixed, lest it encroach into the space over the controls or into the backseat area. The width of the armrest was limited by the space between the seats, which itself was not great, and within this given width a latch mechanism to keep the cover closed also had to be accommodated. In fact, nothing about the overall size of the armrest cover could be changed without a concatenation of other design changes, something no car manufacturer would want to see happen. If the stored cup holder could be fit into a box of given, fixed dimensions, installing a cup holder in an American Volvo would involve only substituting the cover of one armrest/storage compartment with another. Even an older model might be retrofitted with a cup holder by employing a conversion kit.

The resulting spring-loaded Volvo cup holder was a clever solution to this severely constrained design problem. However, the space available in the armrest restricted both the diameter of and spacing between the cup rings. It is because these limiting factors forced the rings to be placed so close together that it is difficult to grab one drink without disturbing the second in the holder. Still, my wife and I have grown accustomed to the idiosyncrasies of the crowded cup holders and have used them for years now without any significant incident. (Compromise and accommodation do not end with the design of hardware.)

In an early version of the Volvo cup holder, the drink rested on a platform that looked like a child's swing. The design enabled the device to be compact in height for storage and to be deployed by gravity. The solution was not very sophisticated, however, and it had the disadvantages of being noisy when empty, susceptible to jamming, and having a bottom that might swing out from under a can of soda. The swing bottoms were soon replaced by spring-loaded ones like those in our 1996 Volvo; the overall design of the improved version is a reasonable compromise among conflicting and constricting constraints. Though it is fair to call it a bad design functionally, it is also only fair to say that it is a good and clever design given the limitations faced by the designers. (Another minor complaint is aesthetic, for the cup-holder device is off center; it is located closer to the passenger side in order to allow the driver—but the driver only—ready access to the handle of the emergency brake, over which even a retractable device could not be placed.)

The problem of fashioning a cup holder for a new car model that is being designed from scratch also involves constraints and compromises, of course, but in this case there is more opportunity for the cup-holder design team to negotiate with other design teams, such as the one drawing up plans for the console, for more space or a more favorable location for the device. The original Ford Taurus is famous for having been designed from scratch, but because this was done before cup holders were standard equipment, they were not incorporated into that model. Our 1987 Taurus does not even have a glove compartment with a door that opens flat to make a tray table. Subsequent Taurus models could be fitted with pull-out cup holders in the dashboard, but these blocked the radio in much the same way as our Volvo cup holders block the window and mirror controls on the console. Retrofitted design seldom achieves a fully integrated look or function. Working within the parameters of what cannot easily be changed is seldom the way to an ideal solution, for it can overly constrain choice and force ungainly compromise.

SIX

Design in a Box

THERE WAS A TIME in the nineteenth century when each application for a United States patent had to be accompanied by a model "of convenient size to exhibit advantageously its several parts." Like works of art in a museum, the miniature machines and devices were put on display in the Patent Office in Washington, D.C., where would-be inventors, like aspiring artists studying the old masters, could seek inspiration by browsing through the exhibit. But even though the models were restricted in size to be "not more than twelve inches square," collectively they occupied a lot of space and quickly overtook whatever room was available for them, sometimes being jammed by the hundreds into the display cases.

By the time of the Centennial Exposition, which was held in Philadelphia in 1876 to commemorate the founding of the country, there were "about 175,000 models, jammed into 362 cases, from 400 to 1,600 in each one." (A fire in 1877 would destroy over 76,000 of them.) Storing and displaying the models had become an increasing burden on the Patent Office, so it is no wonder that the law requiring models had been changed in 1870, leaving the requirement of a model to the discretion of the commissioner of patents, who did not cease making them obligatory until 1880. Still, perhaps out of habit or out of a belief that their chance of success would be helped, some inventors continued to provide models with their patent applications well into the twentieth century.

Attempts to get rid of the accumulated models were only partly successful. The Smithsonian Institution took about a thousand of the models associated with famous inventors. Another few thousand were auctioned off, but the remainder, "packed into thousands of oak crates, went back into storage." These models were eventually dispersed throughout the museum world and into the hands of private collectors. There remains no comprehensive inventory of what was once proudly displayed in a "continuous quadrangular gallery a quarter-mile long" in the Old Patent Office Building in Washington, a structure that now houses the National Portrait Gallery.

Where to put and display things of all kinds is a perennial problem. People and institutions of people tend to be collectors and accumulators of whatever it is they deal with. How to arrange all the stuff is itself a problem in design, and it is a problem that is seldom satisfactorily solved for any length of time. In fact, how to fit a collection of books into an allocated space has kept many a librarian busy for an entire career. When things of any kind have to fit together in a certain way in a certain space to perform a certain function, like large cup holders deployable from a small armrest, we have the typical inventor's and engineer's problem of making a machine whose volume is less than that of the sum of its parts.

This most commonly encountered design problem is familiar to everyone, from eager young children to jaded adults who as children never could fit all their toys into their toy box. The problem disguised as a game might once have been marketed as an educational toy under the imaginary name of Blox in a Box. The rules of Blox would have been simple: First, take the cover off a neatly packed box of blocks and turn it upside down, spilling the multicolored and multishaped contents all over the floor. Then put all the blocks back in the box so that they fit just as they did before and so that the cover can be put back on. We encounter variations of this game throughout life: when we get to arrange or rearrange the furniture in our own room or office; when we have to pack our books for a move; when we have to load the car for a vacation; when we have to wrap Christmas presents for shipment overseas; when we have to repack the parts of anything that we want to return to the store.

Sometimes the blocks are designed before the box. In fact, industrial-design specialists often have to create a box—sometimes termed a *housing,* or *shell,* or *cabinet*—just to hold some newly designed contents. The now-ubiquitous microwave oven is an example. The idea for a microwave oven is attributed to Percy Spencer, a designer of radio tubes who was so good at it that it was said that he "could make a working tube out of a sardine can." In 1946, Spencer was occupied with radar devices for the Raytheon Company. One day while testing a magnetron, a tube that powered radar systems, he found that the chocolate bar in his lab coat had melted. He speculated that the heat came from the magnetron's microwaves and tested his hypothesis by putting other food items in their path. Popcorn kernels and eggs placed in the beam exploded. The microwaves were activating the molecules in the water and fat inside the food, and so cooking it from the inside out.

The first microwave oven manufactured by Raytheon was "huge and primitive—essentially a large, liquid-cooled magnetron pointed directly into a metal box." Its name was far from primitive, however, as was its price. The Radarange, which was about the size of a refrigerator, sold for between two and three thousand dollars. Needless to say, the first Radaranges tended to be bought for commercial and institutional kitchens rather than for homes. A smaller version, selling for $1,295, was brought out in the 1950s, but, not surprisingly at that price, it also did not succeed as a consumer appliance. It was only when Japanese engineers reduced the volume of the magnetron power source so that the oven was about the size of a bread box that the new appliance began to be practical, and by 1975 microwave ovens were outselling conventional gas stoves. (I recently saw a bare-bones model of a new microwave oven on sale for $37.95.) Still, though over 90 percent of American kitchens now have a microwave oven, cooking food from the inside out has not displaced cooking it from the outside in, as conventional gas and electric ovens do. It is the traditional method that produces the more desirable taste for most palates.

The size of things is as important to designers as it is to the ultimate consumer. In the mid-1960s, the engineers Jack Kilby, Jerry Merryman, and James Van Tasse, of Texas Instruments, produced the prototype of the inexpensive battery-powered, handheld electronic calculator. At about the same time, the Hewlett-Packard Company developed the first desktop electronic calculator capable of performing a wide variety of scientific and trigonometric operations. The HP-9100 was a great advance over the old mechanical devices, which could be as large as cash registers and as loud as sewing machines. Still, William Hewlett was disappointed that he could not easily carry his thirty-pound electronic device around with him, and so in the early 1970s, he challenged Hewlett-Packard engineers to come up with a version of the calculator that could fit into his shirt pocket. They measured the size of Hewlett's pocket and set to work on miniaturizing the machine. The Japanese had already come out with a handheld electronic calculator, and Texas Instruments had introduced its four-function arithmetic calculator in 1971, but these did not do much beyond the standard arithmetic operations and so were not versatile enough to let scientists and engineers retire their slide rules. But shortly after 1972, when the handheld, shirt pocket–size HP-35 multifunction scientific calculator was introduced by Hewlett-Packard, the slide rule finally did become obsolete.

Today, several calculators can fit into a single shirt pocket. Some are so thin they can be kept in a wallet with credit cards, and others can be worn on the wrist like a watch. But in achieving such miniaturization, designers and users alike have had to make compromises. The small calculator often takes two hands to operate, since the machine is too light to stay put on a desk when its keys are punched. Some keypads, such as those on calculator watches, are so small and the buttons so closely spaced that they have to be operated by means of a stylus. Otherwise, several keys end up being pushed at once because a finger cannot fit in the imaginary box delineating one key's airspace from that of its neighbors.

But long before the move to microminiaturization, the typical shirt pocket–size electronic calculator still had one of the slide rule's disadvantages: It could not print out a record of its calculations the way an old mechanical adding machine could. Hewlett-Packard did make printing calculators, but these were limited by the small number of characters they could reproduce or by the special thermal paper that they required, not to mention the large amount of battery power that they consumed. Handheld calculators and computers could be hooked up by wires and cables to plotting devices and dot-matrix printers, but these tended to be mechanically temperamental and noisy. Quiet printing was the grail.

By the late 1970s, several companies were trying to develop a new technology of ink-jet printing, in which ink would be silently squirted out of tiny nozzles. Canon was among the manufacturers who were pursuing this technology, and its early efforts focused on using a piezo-electric material, one that would allow the shape of the nozzle to be changed when coaxed to do so by an electric current. When a piezo-electric nozzle was subjected to an electronic pulse, it would expand and contract very quickly, thereby first sucking in a tiny amount of ink from a reservoir and then expelling it onto the paper. By controlling and coordinating the electronic pulses and the position of the nozzle, virtually silent printing could be achieved. But it was not easy and did not produce the sharpest of images.

One day in 1977, Ichiro Endo, working on a piezoelectric device for Canon, accidentally let his soldering iron touch the tip of a syringe filled with ink. The liquid inside the needle boiled, and the expanding bubbles forced ink out of the nozzle. Canon's bubble-jet printer was demonstrated as early as 1981 and was for sale commercially in 1985.

Shortly after Endo had his lucky experience, Hewlett-Packard engineers made the same discovery, and they pursued their own ink-jet technology. As this became more and more promising, the calculator division of the company decided to expand its efforts from simply making a print head for calculators to producing a stand-alone printer that would also work with personal computers. In order to proceed with the development, it was necessary to coordinate the efforts of diverse design teams: those working on the machinery to feed the paper; those developing the electronics to communicate with the computer; and those designing the housing (or box) to contain everything, including the controls and the print head, or pen. This made it necessary to know how large the various components would be. Those members of the design team who could not yet say how much space they needed were assigned a limit: "Since the pen's design was just starting to take shape, the printer designers drew an imaginary box of a size that fitted their needs and gave this as a constraint to the pen designers. The entire print head, ink and all, could take up no more than one cubic inch."

Needless to say, this did not make things any easier for the print-head designers. Any space they used to make their nozzle's mechanism larger usually came at the expense of a smaller ink supply. In spite of being boxed in by such constraints, the engineers accepted the challenge, as engineers are wont to do, and proceeded to compromise between the size and performance of the hardware and the size of the ink reservoir.

The HP-2225 ThinkJet printer debuted in 1984, its name coming from the operating principle of "THermal INK JETting out of a nozzle." Since the original design goal was to produce a printer for pocket calculators, the ThinkJet operated on batteries. Before too long, however, the use of the printer with personal computers overwhelmed the market; well before the end of the century, Hewlett-Packard devices had captured a sizable portion of the printer market, and the company's distinctive packing boxes had become familiar in office-supply and consumer-electronics stores everywhere.

In fact, because I have just replaced the print cartridge in my ink-jet printer, I had to open one of these boxes. Like a lot of small but costly items, the cartridge comes in a package that is much larger than might seem necessary. Inside the trapezoidal box is a compartment into which the foil-wrapped plastic-cradled cartridge fits snugly, but there is a lot of empty space around it. It may be that this empty space is designed to cushion the contents if the box is accidentally dropped from the display rack from which it is intended to hang. Or perhaps the overall box shape was not redesigned when its contents were. Or perhaps the box was made larger than needed to thwart shoplifters with pockets of limited capacity. But designs of everything, from printers and printer cartridges to packaging and catalog numbers, are always undergoing change. Indeed, this particular cartridge used to be part number HP-51626a, but now it is simply part number 26. "Easier number same cartridge," it says on the box.

The two small white plastic trays in which the foil-wrapped ink cartridge is cradled appear designed not only to aid the automated packaging process but also to prevent the cartridge from rattling around in the box and provide a second line of defense to protect it from being harmed should the crushproof box be damaged. (The plastic trays are, no doubt, intended to be discarded, but I have recycled one of them to hold paper clips. Such recycling might even be viewed as a trivial form of inventing and designing.)

Like the Hewlett-Packard printer cartridge, everything must fit into a box, if not before it is designed, then after it is made and ready to be shipped. When compact discs were introduced as alternatives to cassette tapes for playing music, a container for the disc also had to be designed. Thus we have the now-familiar jewel box in which compact discs are sold and meant to reside when not being played. Yet the design of the jewel box is a constant irritant to many music lovers as they struggle to open it or extract its contents. When I give engineering students the option of redesigning any consumer product they wish, many choose the CD jewel box as the object of their attention.

Long before the CD even existed, using cassette tapes in automobiles was also a frequently frustrating experience, given that the tapes were prone to unwind on the mechanism behind the player's flap door. With the establishment of the compact disc as the medium of choice for playing music in an automobile, the CD player came to supplement, if not displace, the tape player. Design teams thus had to come up with a device small enough to fit into a radio, which itself had to fit into its allotted space in the dashboard. In some cases, using the space made available by eliminating the cassette player made the problem solvable, but the different dimensions of the two music media and their decoding mechanisms made the problem more than trivial. Thus, some CD changers are located in the glove compartment or trunk.

Designing an entire car, like the Ford Taurus, from scratch presents a much more complex and difficult blocks-in-a-box problem. Although cars themselves are not shipped in boxes, they must fit in various imaginary boxes, whose dimensions are limited by the width of highway lanes, the height of garage doors, and the length of parking spaces.

However, no carmaker seems to want to make or try to market a perfectly rectangular vehicle (though some sport-utility vehicles and minivans appear to be approaching that geometry). So the size of the passenger, engine, and trunk compartments must be played off against one another as well as against styling and accessory considerations. The radio must fit into the dashboard and the speakers into the door. Putting all the parts together necessarily involves a lot of compromises and trade-offs between the size of the radio and that of the glove compartment, and a lot of arranging and rearranging of parts must take place during the total design process. Yet, in the end, like a box of blocks successfully repacked, the design can appear to be seamless, as it should.

It is a constant challenge to all designers to fit parts into boxes that have fixed dimensions. Often the boxes are metaphorical, but they are constraining nonetheless. In the process of compromising between fitting these parts into this box and those parts into that, and the boxes into each other, the designer—whether architect, engineer, or layperson—must make choices and thus make concessions to the reality of the problem. And it is a problem that we all face at one time or another, especially when we travel.

It seems that almost all luggage boxes, known as suitcases, are black these days, and so they are easily confused at the airport. We have learned to heed the signs in baggage-claim areas that warn us that many bags look alike. It could be embarrassing, to say the least, to arrive in a hotel room late at night and discover that the suitcase on the bed was packed for and by someone of the other gender. Whether the error can be sorted out before the important business meeting the next morning might depend on whether the other person had yet made the same discovery and reported it to the airline in time to make a difference.

In part, fashion and competition are to blame for so many suitcases looking alike. People want to own what is fashionable, and many manufacturers want to ride the coattails of a successful fashion trend. We don't buy luggage solely for looks, however; the popular roll-on suitcase is both convenient and functional. Frequent fliers consider its carry-on size a definite plus, and the wheels and extensible pulling handle are a godsend to a weary traveler. But packing some of the smaller suitcases can be an exercise in frustration precisely because of the many design compromises that need to be made to ensure that the size of the black box conforms to airline regulations.

A chrome box frame into which a piece of luggage must fit can often be found at airline check-in counters beside the entrance to the Jetway. If your bag doesn't fit, you should check it. This truth embarrasses the passenger who does not test his bag in the luggage gauge and then encounters the impossible problem when the bag truly does not fit under the seat or in the overhead bin. Luggage manufacturers have been increasingly attentive to what fits into the chrome box. Since the wheels and collapsible handle must also fit within the box, this means they encroach on the space that might otherwise be used for clothes and travel paraphernalia. My roll-on has a false bottom, into which I can pack socks and other small soft objects around the tubes into which the handle slides, but the limited vertical space fills up quickly. When I add my shaving kit, books, and assorted electronic devices, there is little room left for clothes.

Packing a suitcase has always been a design problem, but with the growing popularity of carry-on luggage, it has become an especially constrained one. The choices and compromises that the manufacturer has to make in fitting wheels and handle rods within the box force the traveler to make more choices and compromises in packing, and these decisions usually involve figuring out how much can be comfortably transported by being worn rather than packed. I have found that my suit does not really fit into my one-suiter and arrive unwrinkled, and so I often wear my best clothes for the trip on the plane. Because I am wearing my suit, I usually pack an extra pair of trousers and an extra shirt or two, in the event that I or someone else spills something on me.

Like Russian dolls, boxes must often be packed inside larger boxes: Suitcases, for example, frequently have to be fit into automobile trunks. Loading an automobile trunk for a long trip presents a tremendous design challenge, and one that is not always met. And so, especially during vacation time, we see cars with luggage racks and clamshell carriers strapped to their roofs. My wife and I prefer to fit everything within the automobile, and so we have rejected car models on the basis of their trunk design. It is not just the cubic feet of space that matters to us; it is also the configuration of that space and the access to it. Our Volvo does not have a particularly large trunk, but, like the car itself, it is squarish in shape and so lets us feel that we are using the space efficiently to contain our squarish luggage.

Unfortunately, the opening to our Volvo's trunk is not very wide at the back or very deep at the top, and so we cannot fit extremely large things into it. We once bought a television set that came in a rather bulky box. We managed to get it home by letting the box hang out the back of the trunk, but on every hill we climbed, we worried that the thing was going to topple out. On another occasion, when we bought an even larger television, which came in an even larger box, there was no hope of fitting even one corner of it into the trunk. The attendant who carted the box out to the curb had seen the problem before and told us that if we took the set out of its box, we could probably fit it in the front seat, if the seat were reclined and pushed all the way back. He was right, naturally, and so we drove home with a new television set in the passenger seat and my wife in the rear seat, steadying the TV around turns.

Though we did not lower the back of the seat fully for the television, the seat does recline into an almost flat position. The backs of the rear seats of our Volvo also fold forward, allowing them to lie almost flat atop the seat bottoms, giving a clear opening between the interior of the car and the trunk. If we choose not to lower the entire backseat, there is a pass-through hidden behind the armrest, allowing skis and similarly proportioned objects to be transported. We have used this feature to transport lumber rather efficiently. With the seats down all the way, we have also carried a sixteen-foot extension ladder (collapsed, of course) wholly within the car, and have transported from the nursery an eight-foot-tall bagged young birch tree lying on its side, with the trunk door closed against the teeming rain.

Mostly, however, we use the Volvo's luggage space to carry bags. Arranging everything in the trunk efficiently is always a challenge, but one that I enjoy. Perhaps all engineers delight in the challenge of packing a car trunk. A woman once told me about growing up in a family of five children; her engineer father may have taken the challenge to extremes. On vacations, each child was responsible for packing his or her own luggage, but only the father was allowed to put the bags in the trunk. Before setting out on a trip, he would pack and unpack and repack the space until everything fit to his satisfaction. The bags were then tagged to identify their place. Each morning, the children were responsible for carrying their own luggage to the car and placing it on the ground, arranged according to its place in the trunk. The father would then pack the car for the day. I don't recall being so obsessive on family vacations, but my wife and children may remember differently. Regardless of how any of us approaches packing a car trunk, the automobile does present a wealth of design problems and solutions, for engineers and nonengineers alike.

"Thinking outside the box" is a popular catchphrase for creative activity. But it does not always work in design, where the box not only confines but also defines the problem. We must also "think within the box." As with a suitcase, we can only pack so much into a design-problem box. What we include and what we leave out are often tough decisions, but they have to be made. Just as travelers have to anticipate what the weather will be like in the location to which they are traveling, so the designer has to anticipate what the box will experience as it is transported from factory to store to consumer. Packing too much into the box can ruin it and its contents; packing too little can cause the contents to rattle and beat against the sides of the box. The trick to packing and design alike is to get it just right, to think within the box as well as outside it.

Sometimes, as desirable as it might be functionally, we do not want to fit everything into a single box, lest the box be too heavy to lift or too awkward to carry. The notebook computer at which I am now working is wonderfully slim and lightweight. To achieve this condition, it was designed without some of the conveniences that my old desktop computer had. My laptop has no built-in floppy-disk drive, no built-in CD-ROM drive, and I can count on its relatively lightweight battery for only two or three hours before it has to be recharged. As a result, I must carry a separate heavy and cumbersome alternating-current adapter when I travel with my computer. But I am happy with my laptop. It has a full-size keyboard, a large and clear display, plenty of internal memory, and software features I am sure I have yet to discover. Though I have given up some things in choosing this computer, I am happy that it does not add excessive weight to or take up too much precious space in my luggage.

When I work on this computer at home, however, I can attach to it an external floppy-disk drive, an external CD-ROM drive, a zip drive, a scanner, a digital camera, a printer, an external power source, a phone line, and other accessories on my desk. The computer may not look as sleek or as portable as it does going through the X-ray machine at the airport security station, but when I work at my laptop, I am looking at the screen and not at the umbilical cords that project from three sides of the base. The thought of all this external equipment disturbed me when the computer was new, but it no longer bothers me. I have adapted to this machine that thinks outside its box.

Design problems and their good-faith solutions are everywhere. The problems may start out little differently than a mess of blocks on the floor and an empty box in the hand, but their solutions are more than child's play. The challenge to the designer is, and always will be, to pick up the blocks one by one, modify them if need be, and arrange them in the box not only so they fit but also so they fit in a meaningful way. When the blocks are numerous and the box is large and heavy, or when there are several boxes that must fit into still-larger ones, it often takes a team of designers or engineers to accomplish that end. There is usually no telling at the outset how that end will be accomplished, but there is seldom any doubt that it can be, as long as the blocks and boxes are not off-limits in terms of being redesigned. Just as any square peg can be fitted into any round hole by making the peg smaller or the hole larger, so any number of blocks can be fitted into any box by adjusting the size of one or the other, or, in the spirit of compromise, of both.

SEVEN

Labyrinthine Design

IT IS NOT only tangible things like lightbulbs and mousetraps that are invented and patented. How we do things has also become the object of patent applications, leading to surprising results at times. Something as old and as simple as playing on a swing might not appear to most adults to be fertile ground for invention. Youngsters everywhere have discovered and invented, and rediscovered and reinvented, and designed and redesigned all sorts of unconventional ways to enjoy themselves on swing sets, including simulating dogfights and moving sideways on them to crash into neighboring playmates. Few have thought to apply for a patent, but not long ago, when one father observed his young boy moving his swing sideways by pushing alternately left and right on the supporting chains, he saw not play but intellectual property. The father, who happened to be a patent attorney, filed an application in the name of his son, the inventor, and the five-year-old was in time granted a patent for a "method of swinging on a swing."

Processes and procedures that involve a clever new way of doing a familiar old thing, or a simple way of doing a complicated series of things, are also designs. Thus, the way we check our bank-account balance or order a product on the World Wide Web involves design, in that decisions have to be made about the way sensitive personal and preferential data is to be used and protected by a bank or a dot-com company. One-click ordering, the "method and system for placing a purchase order via a communications network" introduced by Amazon.com as an expedited checkout system for buying books on the Internet, was, to the disappointment of competitors, declared a patentable way of doing business. This controversial and debated decision became symbolic of the broad interpretation of what has come to be considered a patentable invention in the electronic age.

The idea of designing a shopping experience did not have to wait for the appearance of the World Wide Web. Consider the problem of organizing the contents of a physical supermarket. How do you arrange fifty thousand items, which come in a wide variety of shapes, sizes, and categories, in a boxlike volume containing maybe fifty thousand square feet of floor area? Obviously, it is not just a question of packing the items into the space inside the building, the way luggage is packed into the trunk of a car or blocks thrown into a toy chest. Each of the items for sale in a supermarket must be displayed in its own space, within sight and reach of the shopper of average stature, which means that shelves cannot be too high or too deep. Otherwise, a short shopper has to climb the shelves as if they were a ladder or ask a taller shopper for help to reach the last box of Minute Rice sitting at the back of a top shelf in a poorly stocked or depleted supermarket. To such physical constraints must be added the more intellectual one of arranging the items in a logical way.

It is not just the grocery items that must have their dedicated space. For a supermarket to accommodate customers with some degree of ease of movement, there must remain ample unoccupied space—the aisles—wide enough for two-way shopping-cart traffic, U-turns, and shelfside parallel parking. Thus, there has to be a compromise between cramming as many items as possible within the space of the store and leaving enough room for people to shop. At the same time, because a supermarket is a super business, the management would like to see the wide variety of products arranged in such a way that shoppers will likely leave the store with significantly more products than were on their shopping lists.

From the store owner or manager's point of view, the perfect layout for a supermarket might be a very long, very narrow building with but a single aisle (or, better yet, a single shelf) running the entire length of the store, with the only entrance on one end and the only checkout counters on the other. In such a store, the customer would have to pass by every single item on display and so be reminded or tempted to buy more than is on a shopping list. A building ten feet wide and a mile long would have the requisite fifty thousand square feet and so should suffice for this ideal, though that would mean that each visit to the supermarket would involve a two-mile walk for the customer, no matter where the car was parked. Turning around a short way down the aisle to go directly to the checkout counter in order to make a hasty exit would not be an option in the mile-long store, which might be dubbed the Milemarket, Milemart, or MileAisle. Shoppers having the box of cake mix that they came for could only forge ahead toward the end of the aisle, where it would widen like a toll plaza to accommodate the store's checkout counters.

The apparent marketing problem of requiring the customer to walk two miles for a carton of cigarettes or a six-pack of beer might be overcome by promoting the MileAisle shopping experience as an opportunity to exercise. Unlike mall walkers, who segregate their exercise from their shopping activities, the MileAisle walker/shopper would accomplish the two things at once. Furthermore, because the entire circuit is two miles long, it would not be possible to curtail the distance walked by reducing the number of laps. Once the point of no return was reached, the walker/shopper would have to complete a full two miles of exercise. To attract customers even during rainy weather, a colonnade could be constructed along each side of the building. Such a feature might even increase business during inclement weather.

The problem of replenishing the shelves of a MileAisle store could be solved by widening the building a few feet on each side to incorporate long, narrow stockrooms behind each set of shelves. The shelves could be restocked from behind, the way refrigerated coolers are in convenience stores, thus eliminating the problem of stock boys obstructing customers in the aisle. Even without stock boys in the single-aisle store, however, an enormous traffic jam would, no doubt, result if just a few hypertensive shoppers stopped at the frozen-food section to check for monosodium glutamate in the list of ingredients on a frozen entrée. There might have to be a moving walkway down the center of the aisle, much like the moving walkways in airports, so shoppers in a hurry could speed past the dawdlers. The moving walkway could not have rails, though, because shoppers would have to be able to get on and off it at will and move freely from the shelves on one side of the aisle to those on the other. There could even be an express aisle along the entire length of the store, to accommodate those shoppers who wish to power-walk.

Securing a mile-long but narrow plot of land might be among the greatest challenges to entrepreneurs wishing to establish a MileAisle in a developed area. The single-aisle concept could still be accomplished by a reconfiguration of the design. Instead of a linear store, an annular one could be designed, with a single entrance and exit and strictly enforced one-way circulation. But a circular aisle one mile in circumference would require a plot of land about seventeen hundred feet square, or one large enough to hold a good number of football fields, which could still be a real-estate challenge. To reduce the footprint of a MileAisle store, it could be constructed as a helix, with the shoppers taking an elevator up to the top and walking down, much as museumgoers do in New York's Guggenheim. If local building codes did not allow a tall enough structure, the aisle might be designed as a double helix, with shoppers walking up one strand and down the other, thus also providing a more rigorous exercise regimen. No matter how configured, the single superlong, superrational supermarket aisle, though a manager's dream and likely even a patentable one, might be a risky business venture.

From the shopper's point of view, the perfect layout for a supermarket might be a circular store with a rotating floor, much like that of a revolving restaurant. A 50,000-square-foot store would have a reasonable radius of about 125 feet and a circumference of about 800 feet. If it rotated once every fifteen minutes, the edge would be moving at a bit more than half a mile per hour. It would be no more difficult to enter and exit the revolving SuperCirclemart than it is to step onto and off of an escalator or a moving walkway or a starting or stopping carousel.

(London's Millennium Wheel, the largest Ferris wheel in the world, does not stop turning while its passengers enter or exit its cars, which move at about half a mile per hour.)

The shelves and aisles in the SuperCirclemart would be arranged radially, making them wedge-shaped. The bulkiest, heaviest, and most frequently needed grocery items, like milk, would be located closest to the outside of the rotating floor, thus minimizing the distance they would have to be carried to an exit. The smallest and least frequently bought items, like spices, would be at the center, where the shelves and aisles would converge to minimal dimensions. The stockroom could be in the stationary basement, with cartons of new stock fed up through a central elevator, much like water rising in a fountain. Even with this obstruction at the hub of the store, a shopper could get from any one radial aisle to any other radial aisle with a minimum of walking. In an alternate design, the center of the store would be a totally open space, so that people with shopping carts could crisscross it like pedestrians in a large Italian piazza.

Shoppers would enter the SuperCirclemart at many different points on its periphery, the way children mount a merry-go-round when seeking their favorite horse. As shoppers would drive into and around the SuperCirclemart's annular parking lot, they could see through the store's periphery of glass windows and doors and spot exactly where the aisle they wanted was, or was going to be by the time they parked their car nearest the closest door. Practiced shoppers could position themselves to lead the rotating aisle just right, so that they could park to minimize the walking distance to it. Those dropping by to pick up just one bag of cat or dog food might be able to run in and out before the store moved even an aisle or two beyond their car. Shoppers buying larger orders could time their arrival at a checkout counter, one of which would be located at each of the many exits spaced all around the periphery of the store, so that they could leave the store just as it was rotating past the place where their car was parked.

Checkout counters would be aligned with the outer edge of the SuperCirclemart, and customers waiting to check out would line up parallel to the store's circumference. In this way, they would not block shoppers moving from around the end of one aisle into another, something that is a major source of frustration and annoyance in conventional rectangular supermarkets. On rainy days, shoppers could wait inside or walk around the edge of the store until they came to an exit that was close to their car in the parking lot.

Though the layout of a SuperCirclemart might be the customer's ideal, the configuration would also hold great advantages for the management. The manager's and customer-service offices could be located atop a central hub structure, thus providing unobstructed views down each of the aisles. Customers seeking help finding something could be pointed directly to it from the crow's nest. Potential shoplifters would have nothing behind which to hide.

There is a great deal of room for compromise between the supermarket owner's ideal of a single long aisle and the shopper's dream of a carousel-type store. The actual layout of most real supermarkets is, of course, rectangular, tending toward the squarish, a clear compromise between the competing ideals. Given the box, the basic problem for the supermarket designer would then be to arrange the display cabinets and shelves and items in and on them in such a way that the shopper would be required to move up and down as many of the aisles as possible, even when picking up only a few staples.

Most supermarket layouts are designed with the produce department just inside the entrance. This is not because the management wants to make it easy to pick up a cucumber for that evening's salad, but because colorful arrangements of succulent fresh vegetables are believed to convey a good first impression about the store and imply the freshness of its entire stock. The contents of the produce department also can suggest what might be made for dinner that evening—potatoes perhaps promoting steak sales, and strawberries contributing to the purchase of shortcake and cream—thus giving the shopper a purpose to range deeper into the store, where the meat, bakery, and dairy sections are, and thereby passing enticing displays of sale items of a totally unrelated nature.

Supermarket managers can also get shoppers to pass through more of the store than they might want to by locating commonly needed items far from one another and in remote areas. Thus, in a familiar store layout, milk and eggs are located diagonally opposite the entrance, often in a distant corner, with bread and rolls down the last aisle. In fact, such items are almost always arranged around the periphery of the store, so that the shopper seeking them must pass as many aisles as possible, being reminded or enticed to go down those aisles and buy additional items.

As advantageous as the customary layout is to the store, it can be annoying to the shopper. The fruits and vegetables, being put into the shopping basket first, have to be repositioned or sheltered frequently if they are not to be crushed by the heavy bottles and cans added to the cart in the store's center aisles. Who wants ripe tomatoes or avocados or bananas squashed under a large box of detergent? In some stores, the frozen food is closer to the vegetable department than it is to the milk, so on a warm day, the ice cream might begin to melt before the shopper reaches the checkout line. Such annoyances can be avoided by negotiating the aisles in a manner contrary to the store's design and in a direction against that of most shoppers, but ending up with the bread, eggs, and tomatoes on top is easier said than done. The path of an independent and logical shopper might look something like a drawing done with the continuous line on an Etch-A-Sketch, a constrained design problem of still another kind. But such paths usually require retracing and backtracking and passing some displays more than once, thus giving the store a second and third chance to entice the customer into buying still more.

A modest-size independent market that we began frequenting recently announced that it would be moving from its present five-thousand-square-foot location into a modernized one twice that size. The new store, moreover, will take "a common sense approach" to its arrangement. Recognizing that for many shoppers vegetables represent not the first but the "last stop in meal planning," the produce department will be at the end of the journey to the cash registers. A "main meal" area will greet customers coming into the store, containing "prepared foods and meats, an expanded deli, full service meat counter" and a seafood section. Of course, the store still hopes that shoppers will leave with more than they had planned.

The shopper can counter any supermarket owner's strategy by studying the layout of the store and designing a foray into it. Rather than going up and down aisles serially, the prepared shopper can come armed with a list of needed items arranged in order of the most efficient way to traverse the aisles so that the vegetables, bread, and frozen food end up on top of the pile in the basket. (Preplanning such an itinerary is in itself no simple design problem. In mathematical circles, it goes by the name of "the traveling salesman problem," and when a long shopping or stopping list is involved, it is a classically difficult one to solve with any degree of optimality.) Shoppers might have a relatively easy time planning their itemized itinerary if supermarkets made available maps of their layout or indexes of their aisles, but such is not common practice. Unlike department stores, which display directories, most supermarkets are uncharted territories. Management seems to prefer that a customer roam the aisles and retrace paths already taken to find the canned yams, for who knows what impulse purchases might be made along the route? Managers also seem to fear that regular customers will create their own mental maps after a while, thus going directly to the shelf holding the one item for which they came. Perhaps it is to frustrate such shopper efficiency that the layout of supermarkets seems to be rearranged at regular intervals.

No matter what the internal layout of an actual supermarket is, the checkout lanes are always located near the exit, which may or may not be close to the entrance, as many shoppers note when parking their car. Approaching the checkout, the shopper is presented with a new design problem: how to minimize the time spent waiting in line to have what is in the shopping cart scanned, to have the bags packed, to pay the bill, and to leave the store, only then to have to find the car, and, if the trip has been for a major restocking of the pantry, to fit all the groceries into the car. It is seldom the case, of course, that the shortest line is necessarily the fastest.

What line to choose is another problem in optimization, and it takes experience, careful observation, skillful jockeying, and luck to end up in the line that does indeed move fastest. The regular and calculating shopper will know which clerk has fast hands, knows and remembers the names of unusual fruits and vegetables, and knows that an item that the computer says costs fifty-nine cents is indeed on sale for forty-nine. The shopper seeking to minimize line time must also observe not only how many people are in which checker's queue but also how full their baskets are and what they contain, and whether or not the lane has someone bagging the orders. It is risky to get behind a cart full of beer and wine when the clerk is a teenager and thus will have to call a supervisor over to ring up the sale. A checkout line can also be slowed by a talkative clerk, a troublesome credit card, or a neophyte customer using the electronic credit/debit card reader. Some practiced shoppers choose a checkout lane beside an empty one, prepared to jump over to it as soon as it might open to alleviate the backup. In the final analysis, the shopper must rely also on shopper and checker profiling, instinct, strategy, and luck to proceed through the checkout process expeditiously.

Most supermarkets have express lanes, a feature designed to minimize the frustration of a customer buying a single bottle of catsup by not requiring her to wait behind shoppers with weekly orders so large that they cannot even fully empty their tandem shopping baskets until the conveyor belt moves. Here again, the design concept is only as good as its execution. Prominent signs always state quite clearly how many items qualify an order for express checkout and usually whether or not credit cards and checks are accepted. The question of whether a customer with twelve items can get away with using a ten-item express lane is akin to that of whether a highway driver can go sixty-five in a fifty-five-mile-per-hour zone. It depends on who is on duty that day and on how strictly the limit is being enforced. Whenever it happens that someone with fifteen items wants to pay by check in a ten-item cash line, there are unhappy shoppers watching an imaginary might-have-been space in the next checkout lane, where a customer would have reached the clerk faster. If the express-checkout clerk challenges the presumptuous customer, an ensuing argument could delay things even more.

Supermarket managers and designers are sensitive to such frustrations and realize that these problems can drive customers to a competitor's store, where they can negotiate a less capricious maze. To obviate this, among the latest developments in checkout options is allowing shoppers to check themselves out. They can scan and pack their own groceries and pay the machines with cash or a credit card. To thwart would-be cheats and catch errors, a self-service scanning station not only can provide a monitor clerk with a video of the transaction but also can note each scanned item's weight, which is preprogrammed into the computer. If the customer puts an item in the bag without scanning it properly or at all, the computer notices a discrepancy in the weight of the bag and announces that the last item packed should be removed and rescanned.

Supermarkets have evolved quite a bit from the corner grocery store, but they remain far from perfect. Nor can we expect them ever to be, since countless numbers of constraints exist and compromises must be made in the placement of all the boxes, cans, and bags of merchandise into a fifty-thousand-square-foot floor plan. The constraints and compromises are so contradictory and so complex that even a supermarket's supercomputer probably could not be used to solve the design problem in a way that would suit the storeowner and the shopper equally. In the end, at the checkout counter, the store manager has to hope that each customer is reasonably satisfied and complies with the rules of the queues. The shopper has to rely on instinct, judgment calls, and luck in picking the right aisle to go down for the canned artichoke hearts and the right lane to stand in to be checked out.

Supermarkets are not the only places where we must decide which line to join. Similar situations arise in banks, in fast-food restaurants, and at airport check-in counters and security checkpoints, where the lines in the latter can move at an excruciatingly slow pace toward one's goal of getting on a jetliner that will fly at nearly six hundred miles per hour. The use of rails or cordoned-off waiting lines, in which the first to arrive is the first served by the next available clerk, makes the system seem fairer, but customers find themselves feeling like herded cattle waiting in long, snaking lines, repeatedly coming face-to-face with someone just a dozen places ahead.

Among the most familiar bottlenecks is the toll plaza on a parkway or turnpike. Drivers who had been moving along, if illegally, in excess of eighty miles per hour are suddenly brought to a crawl, if not to a complete stop. To describe conditions at a toll plaza as a bottleneck is, in fact, to abuse the metaphor, which is not apt anyway. A bottleneck should connote a narrowing of the route, as the neck of a bottle denotes its most constricted section, through which the flow becomes turbulent but never stagnant. Paradoxically, a toll plaza occurs at a widening of the route, by perhaps a factor of five or so. At toll plazas, two- or three-lane highways widen into seemingly countless ill-defined "lanes" that reverse-funnel into twenty or so tollbooths, bringing cars to a dead stop. The approach to a toll plaza is, in fact, less like a bottleneck than a bottle cap, or, perhaps more accurately, a stoppered carafe. It is the section of the highway just past the tollbooths that is literally the bottleneck, as cars bolting away like thoroughbreds out of the gate are often brought up short as a field of lanes converges back into two.

Years ago, my brother, then a daily commuter on the Garden State Parkway and a frequent user of the New Jersey Turnpike, gave me some sound advice for getting through a toll plaza as quickly as possible: Always head for the rightmost booths, he declared. This seemed counterintuitive to me, for the fast and passing lanes are usually located on the left. To go to the right seemed the long way around, and, besides, the truck lanes were on the far right. I knew that trucks were long and slow, and that they paid complicated tolls. Tandem tractor-trailer trucks especially reminded me of a once-a-month shopper at the supermarket, dragging two loaded shopping carts into a checkout lane, which everyone else then avoided. But a single large transaction no more fazes the toll taker than the monthly order does the veteran supermarket checker and bagger. In fact, the trucks' tolls are usually taken more quickly and efficiently than those of the cars. Truckers are used to paying tolls and seldom fumble for the cash.

With my attention drawn to the rightmost lanes, I noticed that while most cars waited in long lines to reach the designated CARS ONLY tollbooths, which are more or less aligned with the main course of the highway, the truck lanes moved steadily and smoothly. Signs reading CARS ONLY clearly prohibited trucks from using the lanes so marked, but those labeled TRUCKS and WIDE LOADS did not exclude cars. Frequently, in the time it took to detour to the right and head for a tollbooth with a single large truck that had just pulled in, the truck would be pulling out of the booth as I arrived. Why, having had the opportunity to observe such efficiency from the fixed vantage point of a car standing still in one of the leftmost or even one of the rightmost CARS ONLY lanes, more drivers did not head for the extreme outside truck lanes was something I could not fathom. But until my brother pointed it out to me, I had not noticed the advantage either.

Though most drivers may not care to analyze the process fully enough to optimize their passage through toll plazas, this is not to say that any of them enjoy the delays they encounter. It was not too long ago that commuters who had to use toll roads, bridges, and tunnels twice daily, and usually during rush hours, got stuck at the toll plaza coming and going, doubling their frustration and the work of the toll takers. How many commuters might have thought in the morning traffic jam how nice it would be if they could just prepay the return toll and have an unimpeded drive home in the evening? By 1968, the Golden Gate Bridge and Highway District evidently had had essentially the same thought, and the Golden Gate became the first bridge to introduce the practice of collecting tolls only from traffic entering the city. The toll was effectively doubled, of course, but since virtually all of the people driving to work in the morning would be retracing their route over the same bridge in the evening, there could be few objections. (Tourists just passing through San Francisco and thus taking the bridge only one way are either shocked at the high toll or pleasantly surprised at the free passage, depending on the direction of their travel.) The idea of collecting tolls only one way, effectively prepaying for the return trip, soon became common practice. The amount of time, fuel, and frustration saved collectively has been enormous.

Tollbooths were still tollbooths, however, and each car had to come to a stop so that its driver could hand the money to an attendant or toss it into a hopper. After doing so, the second or two it took the gate to rise or the light to turn green seemed to be longer than it actually was. Those commuters who could do so chose roundabout toll-free routes, but most knew of no alternative route that was faster, even with the delays at the toll plaza. Highway, tunnel, and bridge authorities, which saw the lost commuter as lost revenue and also saw the wages of every toll collector as a liability, looked for still more efficient ways to collect tolls.

In the late twentieth century, the accelerated development of microelectronic devices of all kinds held out the possibility of collecting tolls remotely as vehicles passed through a toll plaza without stopping. If the vehicle were fitted with a transmitter/responder (a "transponder") that could transmit a unique identification number in response to a prompting from a radio signal sent from the toll plaza, then the amount due could be deducted from a prepaid account. This is essentially the principle behind automated toll systems, which go by different names in different states and consortia of states. In Massachusetts, the system is aptly and brilliantly called Fast Lane; in Maine, it has the more pedestrian name of Transpass; and in New York, New Jersey, and several other states, it goes by the typographically challenging E-ZPass.

Technically, such systems work like a charm. The transponder is a black box (though E-ZPass's is actually white), not much larger than a deck of cards. The box is usually placed on the dashboard or attached via Velcro-like strips to the portion of windshield behind the rearview mirror. (Some car models have windshields whose type of glass interferes with the signal, and on these, a transponder must be mounted on the outside of the car.) Each time the car and its transponder pass through a toll-collection point, the amount of the toll, which in some cases is discounted to encourage the use of the system, is deducted from a debit account established with the toll authority. On turnpikes that issue tickets on entering and collect tolls only on exiting, the computerized system notes the entrance location and deducts the appropriate amount when the vehicle exits the road, without it ever once having to stop. If credit has been established, the computer will automatically replenish the transponder account by charging the credit card when the balance reaches a preset low level. Users are regularly sent statements detailing transaction activity, and in the most sophisticated systems, the computer "learns" the travel patterns of users and adjusts their account balances according to projected needs. For those who choose not to give a credit-card number to a computer, accounts can be replenished with cash.

Ironically, in New Jersey it was not the accounts of E-ZPass holders but those of the E-ZPass system itself that began to run out of money. When that state set up its automated toll-collection system, it calculated that there would be a certain number of drivers who would try to escape paying tolls altogether by speeding through unmanned E-ZPass lanes without a transponder. To recover the lost revenue and more, it was planned that the license plate of a violator's car would be photographed and then the owner would be fined appropriately. The additional revenue collected by this means was projected to help pay off the debt incurred in installing the electronic system. After a year or so of operation, however, New Jersey found that it was spending more money to collect from the scofflaws than it was bringing in. To make up for the loss, it was proposed that New Jersey E-ZPass users be required to pay a per-month service charge, something not done in any other state in the consortium.

Speeding through a tollbooth, with or without a valid transponder, is also a concern to those who man the booths. In many cases, vehicles are advised to slow down to as little as five miles per hour when passing through an automated toll lane, which is usually only about ten feet wide. Such low speed limits may not be necessary for the electronics to work, since it has been reported that a toll can be collected from a car going as fast as eighty miles per hour. Rather, the speed limits have been put in effect for the safety of other motorists in the area of the toll plaza and of the toll takers, who have to walk across lanes of traffic when changing shifts.

The speeding problem got to be so bad in New Jersey that drivers were threatened with losing their electronic privileges if they passed through the lanes too quickly. Though there are posted speed limits at tollbooths, drivers are typically allowed a certain unstated leeway, as they usually are on the open highway. In New Jersey, those cars that exceeded the real limit were sent warning letters, something that is easy to do with the computerized system. After two warnings, a third tollbooth speeding violation brought a thirty-day suspension of E-ZPass privileges throughout the states in which they were effective, and a fourth brought a sixty-day suspension. Most drivers are willing to slow down to five or ten miles per hour because of the great benefit of not having to come to a complete stop, open the window, pay a toll with cash, wait for the signal to change or the gate to rise, and generally wait in long lines with slow attendants or recalcitrant coin hoppers.

Toll authorities like a system such as E-ZPass because of its efficiency. A fast human toll taker, assisted by cooperative drivers, might be able to collect maybe five tolls a minute, or three hundred tolls per hour, thus taking an average of twelve seconds per car. In comparison, an E-ZPass lane can reduce the average time between cars to as little as three seconds, thus handling twelve hundred cars per hour. At the posted speed limit of five miles per hour, this would mean that the vehicles were spaced about a car length apart. This may be a reasonable safe-driving distance, but even E-ZPass tollbooths are seldom utilized with totally uniform spacing, suggesting that the average speed of cars passing through the automated lines must be significantly greater than five miles per hour to achieve the efficiency desired by toll authorities and commuters alike. No wonder the five-mile-per-hour speed limit appears to be a mere suggestion.

As the popularity of E-ZPass and similar systems has grown, their smooth operation has been limited largely by the number and location of booths equipped to handle electronic transactions. From the point of view of the driver using an unfamiliar road, the biggest frustration in using E-ZPass is knowing where the automated lane or lanes will be located from toll plaza to toll plaza. On many highways, there seems to be no pattern, and there certainly is no consistency from highway district to highway district. Like the cash-paying driver and the supermarket shopper, as each toll plaza is approached, the E-ZPass user has to survey the expanse of congested lanes ahead and make a split-second decision about which one to head for. Where I have driven in Maine, the automated Transpass lane always seems to be located in the rightmost position, making it easy for users to drive directly to it, sometimes in a dedicated approach lane that prohibits non-passholders from it for a mile or so ahead of the toll plaza.

Wherever the automated lane or lanes are interspersed among the conventional tollbooth lanes, the chance of being able to drive directly to them and pass through even at five or ten miles per hour is severely reduced in heavy traffic. Especially during vacation time or holidays, when so many drivers are strangers to an area, the seeming ambiguity of the various signs above the tollbooth lanes often creates confusion and major lane changing before and across the automated lanes. The ultimate frustration for the E-ZPass user is to end up behind an out-of-state car without a compatible transponder, sitting and waiting for the red light to turn green or for the gate to go up.

In spite of some of the early glitches and annoyances of using E-ZPass or a similar system, the scheme has revolutionized toll collection and will no doubt continue to grow in acceptance and use. In time, the different road authorities that presently use incompatible electronics are likely to establish reciprocal agreements and to cooperate in setting standards that will enable a vacationer with a single transponder to travel from Maine to Florida or from New York to Los Angeles without having to stop at a single tollbooth. This may not occur anytime soon, however. As of late 2000, electronic collection of tolls had been established in only nineteen states, plus Puerto Rico, using twenty different brand names of systems. Florida alone had five different automated toll-collection systems: O-Pass, C Pass, Lee Way, EPass, and SunPass.

The ideal for many supporters of automated toll collection is not only to have a single system but also to eliminate tollbooths entirely. Some toll roads have already removed all tollbooths and special speed limits at toll-collection points. These include a highway outside Toronto and roads in Houston and Denver, where the system is called ExpressToll. Cars and trucks equipped with a transponder similar to that employed by E-ZPass just drive past toll-collection points at normal speed, the overhead electronics handling the accounting. Those highway users without compatible transponders or with no transponder at all have their license plates photographed and then the owners are sent bills in the mail. Similar traffic cameras have, of course, been installed at intersections to record moving violations.

As it is a law of physics that for every action there is an equal and opposite reaction, so it appears to be a law of design that for every invention there is an equal and opposite invention. One such contrary invention, designed to thwart police cameras, has been patented by Peter Kaszczak of Yonkers, New York. He has devised an "ultraviolet laser emitter that is mounted next to a car's license plate." This scrambler of sorts "ensures that the photograph taken by the automatic camera is either inaccurate or blurred, which makes detection of the violation or the violating automobile difficult." Though the use of such a device is likely to be banned from city streets and toll roads the way radar detectors are from state highways, until the inevitable reactive legislation is passed, the invention may attract its share of investors.

Whether surveillance cameras and their scrambling counterparts are compatible with individual freedoms will ultimately be a matter for the courts to decide. In the final analysis, the design of a toll-collection system, like that of a supermarket layout and checkout scheme, must accommodate all constraints by compromising among them and their advocates and detractors, whose concerns are sometimes inconsistent with their actions.

Although electronic systems involving E-ZPass transponders and other identifying devices are completely voluntary, objections have been voiced by civil libertarians and others that electronic systems can also monitor the whereabouts of individuals and can track vehicle travel patterns. There are fears that Big Brother will monitor billions of daily electronic transactions. Perhaps many of those cars waiting to pay cash at crowded toll plazas are driven by people who worry about such things.

I know of no one who refuses to shop at a supermarket that uses lasers to read the zebra stripes on everything from soup to nuts. Indeed, I do not know if it is any longer possible to do so. The automated checkout devices that were so obtrusive when introduced are now virtually invisible to all but the technophile. Indeed, the system has been so overwhelmingly successful that it is being taxed, and by 2005, scanners will have to be able to read not only the present twelve-digit bar codes but also expanded thirteen-digit ones. Yet, every purchase in a supermarket, department store, drugstore, bookstore, or any other kind of store can be fed into a central computer in conjunction with the information on a credit card used to pay for the purchase. Many of us now buy gasoline with a credit card, thus enabling our whereabouts to be noted. In between fill-ups, we visit automated teller machines and make telephone calls. Collectively, such information can be used to keep tabs on us, as some highly visible police investigations have made clear, and the information gathered can be crucial to solving crimes. For those who have something to hide, and thus for those who wish to avoid such transactions, daily life can be very inconvenient, as perhaps it should be. Most of us are willing to compromise our absolute privacy for convenience, and this principle is, in the end, most likely to make automated toll collections on highways everywhere as commonplace as walking through a metal detector. Metal detectors themselves will most likely increasingly become incorporated into the interior design of airports and public buildings, thus putting them out of sight and out of mind of all but those with evil intentions or those who suffer from paranoia.

The technology for intrusive surveillance has long been in place, but that does not mean it is being abused. The overwhelming amount of data being collected is a mitigating factor. Many shoppers affix to their key chain a special-customer discount tag, which they allow to be scanned before their order is rung up, in exchange for discounts on featured items. This makes it easy to collect shopping-habit data on individual customers, including not only what they buy but also when and how frequently they buy it, making it easier for stores to match their stock to the demand for certain items. Shoppers who use such tags do not have to wear license plates or name tags to be identified, nor do they have to have their picture taken, except perhaps to join certain warehouse clubs.

Supermarket checkout lanes will likely someday be as fully automated as highway tollbooths. Grocery shoppers will be able to attach to their key chains a transponder like those for automated toll collection and just roll their fully laden shopping carts straight out the door, the contents having been scanned in place or recorded in some other electronic way and the amount due charged to a credit card or deducted from a debit account previously established at the store. Like the highway authority, supermarket managers may know exactly when we passed through, but what they will really care about is whether what we bought has to be restocked, whether we had been enticed to go down certain aisles in the store, and whether there is enough in our account to pay the bill. The system will be designed with the goal of moving merchandise and collecting money, with bottom lines, not Big Brother, in mind.

EIGHT

Design out of a Paper Bag

AMERICAN SUPERMARKET checkout lanes would be even more frustrating bottlenecks than they are were it not for developments in technology. The Universal Product Code and the decoding laser scanner, introduced in 1974, certainly get a shopper's groceries tallied more quickly and accurately than the old method of inputting each purchase manually into a cash register. But beeping a large order past the scanner would only have led to a faster pileup of cans and boxes down the line, where the bagger works, were it not for the introduction over a century earlier of another technological masterpiece.

Originally, bags for merchandise were made on demand by storekeepers, who cut, folded, and pasted sheets of paper, making versatile containers into which purchases could be put for carrying home. The first paper bags manufactured commercially are said to have been made in Bristol, England, in the 1840s. In 1852, a "Machine for Making Bags of Paper" was patented in America by Francis Wolle, of Bethlehem, Pennsylvania. According to his own description of the machine's operation, "pieces of paper of suitable length are given out from a roll of the required width, cut off from the roll and otherwise suitably cut to the required shape, folded, their edges pasted and lapped, and formed into complete and perfect bags." The "perfect bags" produced at the rate of eighteen hundred per hour by Wolle's machine were, of course, not perfect, nor was his machine. He patented a more elaborate, and presumably improved, "Machine for Making Paper Bags" in 1855. However, the operation of both machines must have been plagued by being jammed with the loose pieces of paper that resulted when each fresh sheet was "cut to the required shape," for Wolle patented a third machine in 1858, among whose "novel arrangements" was a provision "for preventing the loss of the strips of paper usually cut off in order to make the bottom lap or seam of the bag." In other words, his new, improved machine left an extra flap of paper at the bottom of the bag, out of sight and out of the mind and guts of the machine.

Early paper bags, like those made by Wolle's machines, had what is known as an envelope bottom, so-called because of what their shape resembled and how they were formed. Indeed, the drawing sheets of his second patent are headed "Paper Bag & Envelope Mach." Though the envelope-bottom bags that all of Wolle's machines produced had the advantage of relatively simple assembly, the product also had severe limitations. In particular, an envelope-bottomed bag could not stand upright by itself and so had to be held open with one hand while being filled. Furthermore, such a bag could not easily accommodate large, bulky items, like hardware goods and groceries. There was thus plenty of room for improvement in the design of paper bags.

The way Wolle's machines formed an envelope bag was akin to but not quite as elaborate or complete as the way an aerogram is prepared for mailing, with its several tablike edges folded over and pasted down to make a compact flat package. There is never a single way to design anything, however, and just as there are alternate ways to fold a piece of paper into an envelope, so there are alternate ways to make paper bags. One method is to overlap and paste together two opposite edges of a rectangular sheet of paper, thus forming a tube open at both ends. When flattened, one end of the tube can be folded over and pasted to the side, thus forming the bag. The bottoms of padded, manila, and other large mailing envelopes are essentially made in this way, and envelope-bottom paper bags are commonly used to this day by stationery and other stores that sell flat goods. However, the bags for the kinds of goods sold in hardware and grocery stores are formed by closing off the end of a paper tube in an entirely different way.

The invention of the familiar square- or flat-bottomed paper bag—the "grocery bag"—is commonly but incorrectly attributed to Luther Childs Crowell, of Boston, Massachusetts, who in 1872 received a United States Patent for an "Improvement in Paper-Bags." The word *improvement,* which is encountered as frequently in the titles of patents as *mystery* is in those of thrillers, is, in fact, a clear giveaway that Crowell's bag represented an evolutionary rather than a revolutionary invention. Nevertheless, this obvious clue is all too frequently overlooked in the literature of technological whodunits. In fact, because the page of drawings accompanying Crowell's patent is headed simply "Paper Bag," a quick reading of the evidence can lead one to jump to the erroneous conclusion that he was the inventor of the paper bag itself.

Luther Crowell certainly did invent a "new, improved" way of making paper bags, and so he rightly earned a patent for his advancement of the state of the art, but he did not invent the original square-bottomed bag. Indeed, by his own admission, he was "aware that paper-bags have been made which will assume a quadrangular shape when filled." He only claimed his method of making them to be "the most simple and practical," and given "the proper machinery" such bags could be made "as economically and as rapidly as the common bags." Indeed, manufacturing paper bags of all kinds had become a competitive business, and more than one inventor had been working on new ways to produce them more quickly, more efficiently, and more reliably.

One was a creative woman. Margaret E. Knight, who has been called "the most famous 19th-century American woman inventor," was born in York, Maine, in 1838; she grew up and was educated in Manchester, New Hampshire. As a child, she preferred "a jack-knife, a gimlet, and pieces of wood" to dolls. She made playthings for her brothers and became "famous for her kites." Her sleds were the envy of the town's boys. Like many a young girl of her times, Margaret went to work in a cotton mill, where one day she saw a steel-tipped shuttle shoot out of its loom and injure a worker. At twelve years of age, she devised the protective feature of a loom-shuttle restraining device, thus demonstrating her talents for mechanical invention, which she would practice throughout her life.

Knight left the mills in her late teens and engaged in a variety of small temporary jobs and activities, which nevertheless introduced her to a wide range of technologies, including those associated with upholstering, home repairing, daguerreotypy, and engraving. After the Civil War, she worked for the Columbia Paper Bag Company in Springfield, Massachusetts, where she became acquainted with the process of making bags from flat sheets of paper. After a while, she began to experiment with a machine that could feed, cut, and fold the paper automatically and, most important, form the squared bottom of the bag.

It was a square-bottomed bag that could be opened to a wide, flat base that Columbia Paper Bag and Margaret Knight's machine were making around 1870. After testing and refining a wooden prototype of her ingenious invention by making thousands of flat-bottomed paper bags with it, she contracted a Boston machinist to fashion an iron model for submission with a patent application. While that model was being produced, it was seen by an unscrupulous would-be inventor, Charles F. Annan, who applied for a patent in his own name but based on Knight's idea. When Knight learned of Annan's action, she took him to court. Annan apparently was counting on the prejudice of the time that a woman could not be a credible inventor, arguing that "she could not possibly understand the mechanical complexities of the machine" that made paper bags. But Knight's "drawings, paper patterns," and more, including relevant entries in her personal diary, convinced the court of her mechanical aptitude and priority of invention, and it ruled in her favor.

Unlike many a contemporary female inventor and writer, Margaret Knight did not conceal her gender by employing only the initials of her given names. She had used her full name on her very first patent, issued in 1870, which was for an "improvement in paper-feeding machines." This "pneumatic paper-feeder" had applications to printing presses and paper-folding machines, which must have been her principal objective.

While the inventor's name on the heading of the five sheets of patent drawings is given in the androgynous form of M. E. Knight, the bottoms of these same sheets clearly identify the inventor as Margaret E. Knight, as she is also identified on the front page and continuation sheets.

It was her second patent, issued in 1871 for an "improvement in paper-bag machines," that dealt with the satchel-bottom grocery bag. Among the patent drawings is one that clearly shows the rectangular shape and flat bottom of the opened bag to be essentially the same as those of brown paper grocery bags today. Knight's machine worked by pulling from a roll of paper stock a sheet that it immediately started to form into a tube. Paste was applied where one side of the paper overlapped the other, thus completing the tube. Knight's machine performed its greatest magic by shaping the end of the tube into a flat bottom by means of a series of three folds, and the drawings that delineate the three-step mechanical folding process look like instructions for "industrial origami": The first fold formed the end of the tube into a slit diamond, the second creased one tip of the diamond over to make a pentagon, and the third creased the other tip over to form an elongated hexagon. With the proper pasting taking place simultaneously with the folding, the closed bottom was formed quickly. The bag was completed by being severed from the continuously forming tube, at which point the cycle was repeated.

To highlight the key feature of her invention, but perhaps also to underscore her legal victory over Charles Annan, Knight declared in her patent that she believed herself "to be the first to invent a device to hold back or push back a point or portion of one side of the paper tube while the blade or tucking-knife forms the first fold," making no claim to the invention of the paper bag itself. She referred to the patent drawings and, in the legal language of a savvy inventor, emphasized that they did not represent the only way that her ideas could be embodied in a machine that accomplished the essential folding step that was "the basis of the flat-bottomed bag."

With her patent in hand, Knight found a partner in a Newton, Massachusetts, businessman, with whom she established the Eastern

Paper Bag Company in Hartford, Connecticut. Her financial arrangements with Eastern gave her $2,500 outright, plus royalties and company stock. Another of her patents, for an improvement in paper-bag machines, was issued in 1879 and assigned to Eastern. Knight's financial arrangements with the company brought her a comfortable income for the time, but since the royalties were capped at $25,000,

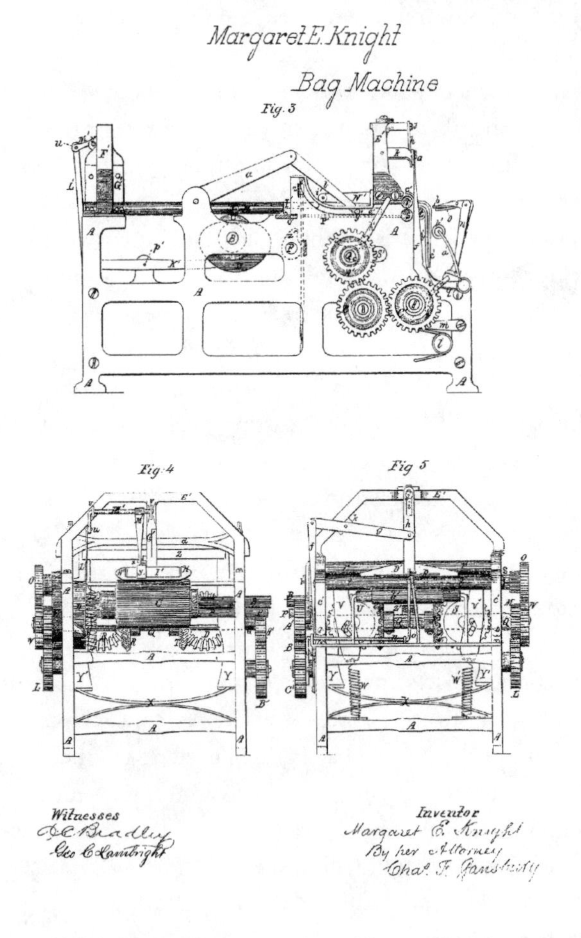

Drawings in an 1871 patent show views of Margaret Knight's paper bag–making machine.

they did not continue indefinitely. Like a lot of inventors driven by the challenge of the new, Knight went on to other things. She eventually received patents for a shoe-sole cutting machine and for improvements in automobile engines.

Though Margaret Knight's flat-bottomed bag could be opened into a boxlike quadrangular shape, it did differ from today's grocery bag in one important way: It did not have the now-familiar accordion-folded

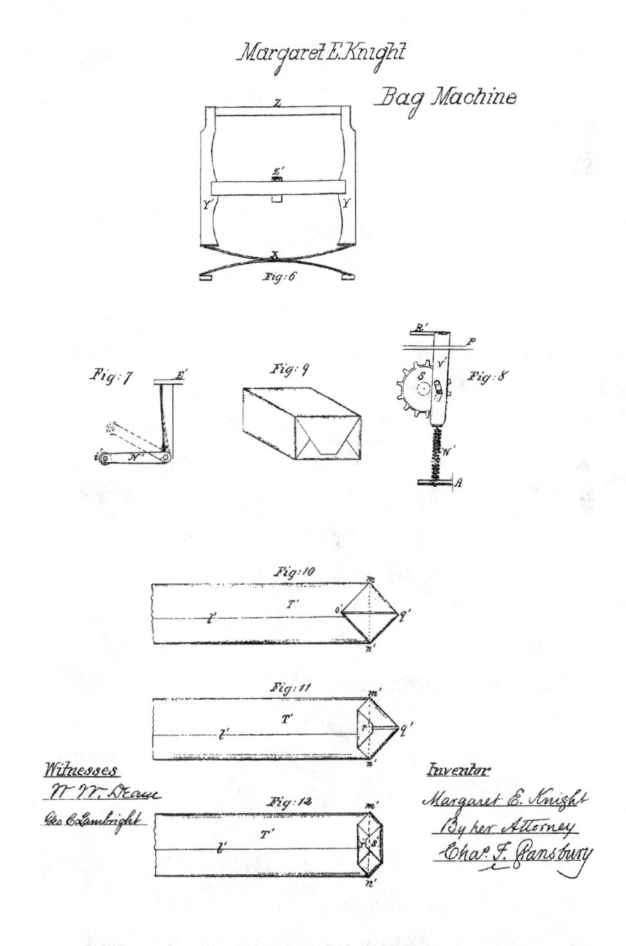

Folding and pasting steps used in forming a flat-bottomed paper bag are illustrated in Margaret Knight's patent.

sides. This reentrant feature not only makes for more compact storage but also defines the corners of the bag by creasing them as part of the forming process. This also enables the bottom of the bag to be shaped in a manner that does not require Knight's ingenious but somewhat difficult-to-manage slit-diamond folding step. Indeed, it was an accordion-pleated bag, with its necessarily different forming folds, that was patented by Luther Crowell in 1872, the year after Knight's first patent for a bag-making machine was issued. Unlike Knight, he did not patent a machine for making his bags, but his could be made more easily by either hand or machine than hers, and hence his claim for his method as "the most simple and practical." However, as with Knight's, Crowell's bag did not unfold easily into a square-bottomed shape. Rather, it had to be coaxed out into the familiar square shape by the packer's hand or by the force of the contents being stuffed into it. But no matter how made or used, the geometry of a paper bag has not lost its magical appeal to those who are fascinated by how even the most ordinary of things are designed and made.

Both Knight's and Crowell's patents understandably focus on forming the bottom of the bag. However, a bag has two ends, and after it is formed, it is not the closed bottom, but the open top, that first commands the attention of the user. A paper bag formed with perfectly congruent top edges, as Knight's appear to have been, can be as frustrating to open as turning the pages of an unread newspaper. The sides cut to the same length tend to stick together and hide their seams like the sides of a plastic garbage bag taken fresh from the box.

Crowell's patent drawings also call attention to the top of his bag, but only because of its irregularity. One drawing shows clearly a bag-length section of a pleated paper tube terminating in the now-familiar and characteristic zigzag pattern, no doubt made by a serrated knife edge but looking as if the cut had been made with pinking shears. More important, though, the drawing shows one side cut a bit shorter than the other, a necessary feature for making the bags according to his improved design. Indeed, after the reentrant folds had been made in a tube of paper, Crowell's bottom-forming process involved only a single folding and pasting, a definite improvement over Knight's.

Designs of all kinds can have little features that may or may not have been intended by their inventor. In the manufacture of Crowell's bag, the tube of paper could not have been simply severed with a single straight knife cut, like Knight's was, for that would have left no tab to fold over and paste to form the bottom. Thus the tube had to be cut to two different lengths. This process left, as an artifact of the manufac-

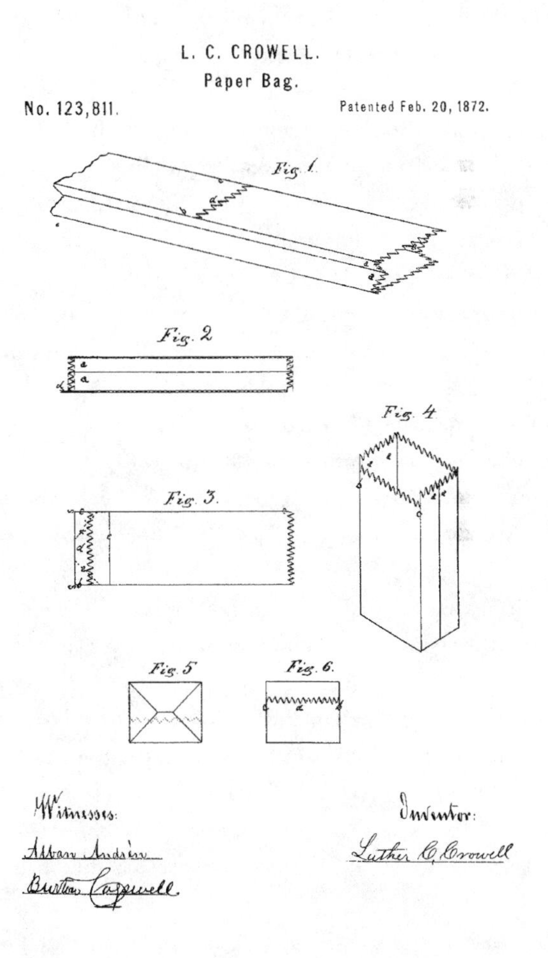

Luther Crowell's 1872 patent illustrated a new way to form a square-bottomed paper bag, complete with accordion-pleated sides.

turing process, a bag top with unequal front and back sides. Rather than being a blemish, the detail proved to be a boon, for the bag could be opened with ease.

Today's paper grocery bags do tend to be formed with both sides of the bag top cut off at the same length, of course, but with a thumbnail or rectangular notch cut into one of them to achieve the same effect as Crowell's perhaps accidental device. (The mating tab on the bottom of the next bag to come down the tube is often folded under in the manufacturing process and so leaves no hint of its presence. Sometimes, however, the tab remains exposed, thus providing a clue to anyone trying to figure out how to reverse-engineer the darn thing.)

As most older shoppers will remember, when a new paper bag was picked up off the pile under the counter in the grocery store, the notch permitted the bagger's thumb and fingers to grasp only one side of the bag. This enabled the bagger in one sweeping motion to snap it open with the flair of a waiter opening a napkin before placing it on a diner's lap. The experienced and flamboyant bagger could open the grocery bag with a loud report as sharp as the crack of a whip or the sound of a home-run ball leaving a baseball bat. No matter how opened, however, the modern flat-bottomed bag stands upright by itself, freeing the bagger's two hands to pack groceries into it with the speed and deftness of a professional juggler. Packing a paper grocery bag quickly and efficiently became a source of pride. It also became the object of competitions of no insignificant skill, given the wide variety of sizes, weights, fragilities, and temperatures of cans, jars, boxes, and other assorted containers that the checkout person passed down the counter to the helper. The practiced bagger was ready for them all, though not always for the likes of David Letterman, who in the 1990s spoke often of packing groceries in his youth. His *Late Show* was the venue for annual head-to-head competitions, though hardly serious ones, between Dave and the current year's championship grocery bagger.

The brown paper grocery bag was not the butt of the jokes, however. The square-bottomed bag itself is a tribute to man and woman's creativity, as manifested in a series of ingenious folds made on a tube shaped from a roll of paper. Inventors and designers like Knight and Crowell fashioned their bags as much with their minds as with their hands. Though these inventions were certainly worthy of having been declared "perfected" even in the nineteenth century, the use to which the bag is put itself represents a recurring problem in design. How do you bag items on a conveyor belt full of ever-changing merchandise? Naturally, harder and heavier things, like cans of vegetables and soup, are properly put in first, providing a solid base that will also give the full bag a low center of gravity. Boxes of macaroni, cake mix, and the like also provide solid foundations on which to pack smaller, lighter, and flimsier items, like packages of Jell-O and plastic bags of beans. Bread and eggs naturally belong on the very top, but not so far up that they can tumble out when the bag is grasped and lifted. As obvious as such protocols might be, they were often violated by inexperienced, careless, distracted, malicious, or mischievous baggers. Whatever experience or practice it may require to pack a grocery bag properly, however, it takes only common sense to recognize an improperly packed bag. As hard as the design of anything may be, fair criticism of it can come from anyone.

The material out of which the familiar grocery bag is made is known as Kraft paper, after the German word denoting power, force, and strength. The name thus connotes the familiar toughness of the bag. Kraft paper is made from a pulping process employing a long-fibered softwood like southern pine. When the paper is unbleached, the familiar brown paper bag results. Bleached Kraft paper is usually used in making bakery bags, the white paper suggesting a cleaner container, one suitable for holding unwrapped rolls, bread, and pastries. Kraft paper bags are also produced in different sizes and weights, with the largest and coarsest bags usually designated as grocery bags.

The strength of a lightweight white bakery bag is seldom tested. It is not intended to hold very much weight, and its contents do not have hard, pointed edges that might puncture or tear through the sack. Brown paper grocery bags, however, can be packed too full and with items that are too heavy, and their contents can have numerous sharp corners that could poke through the paper. The overzealous bagger who tries to squeeze a box of cat food in beside the other boxes already in the bag can often tear a gash along the side. But for the tough structure, this is usually a benign failure, which only prevents the bag's reuse. It is the grocery bag that is overly heavy because of a jumble of cans and jars and bottles inside that poses the more dangerous situation. There is a tendency to pick up a bag by its edges, which can usually be done safely only for a light load. The overly heavy bag has a tendency to rip at the stress-concentration points beneath one's grasping fingers, a situation that, in accordance with Murphy's Law, occurs as the bag is being lifted out of the shopping cart beside one's car trunk. The result can be a mess of glass and jelly on the pavement of the parking lot. Like any design, the so-called perfected grocery bag is only as good as the care with which it is used.

However, it is not the paper bag's susceptibility to failing under extreme conditions that has been its downfall. As every shopper knows, the brown paper bag has become increasingly scarce in supermarkets for reasons other than strength. It has been largely replaced by plastic bags, whose propensity to develop, but also to tolerate, holes and rips and tears is well known to anyone who has brought a supermarket order home. The plastic grocery bag also has a smaller capacity than the paper bag, and its thin plastic film suggests flimsiness and ephemeralness rather than the strength and substance of a Kraft paper bag. So why by 1996 were more than four out of five grocery bags made of plastic?

The plastic bag was introduced into American supermarkets in the mid-1970s. In 1982, only 5 percent of grocery bags were plastic, but increasingly shoppers would be asked, "Paper or plastic?" By 1990, plastic bags accounted for 60 percent of the market, or about 23 billion bags per year, and some stores were no longer giving shoppers a choice. The plastic bag, with its lighter weight and more compact form, was the clear preference of the merchant and, in spite of its shortcomings, in time became the overwhelming choice of the customer.

Plastic bags begin as long seamless tubes extruded from various kinds of polyethylene, a form of plastic that in sufficient thickness can have considerable resistance to stretching, a desirable property for a grocery bag. However, instead of their bottoms being formed from the folding and gluing operations that produce paper bags, individual plastic bags are formed by applying a line of heat across the flattened tube of plastic film. This accomplishes three things: It fuses the polyethylene into an almost-seamless bag bottom; it separates that bag from the next in line on the tube; and it fuses the top of that bag, thus forming the basis for its handles. After the (still-closed) bags have been stacked, they can be cut through to form an opening between a pair of handles.

Because their handles are suggestive of shoulder straps, the plastic bags are known as "T-shirt" bags within the industry. It is a misnomer, however, because when new and still unopened, their handles are suggestive not of a short-sleeved polo shirt but of the shoulder straps of an even more casual muscle shirt. Nevertheless, the bags do have great advantages for the customer. They are easier to wad up and store at home, and as long as they remain intact, their waterproof properties make them excellent for reuse. They can be used to dispose of wet refuse, for example. (Taking out the garbage in an old brown paper grocery bag often results in a trail of drippings that Hansel and Gretel could have followed back to their house.) Plastic grocery bags can also be reused to hold wet laundry, dripping swimming suits, and just about anything, damp or dry, that has to be carried somewhere without making a mess.

It should come as no surprise that the principal reason that plastic has displaced paper in the grocery store is economics, aided and abetted by promotion. Not only were costly paper bags given away free with each grocery order, but they also took up a great amount of space, which consequently could not be used to stock things that people might buy. So it was not just to save trees that some stores began to offer customers a cash credit, albeit usually only a nickel, for bringing their own bags to the store.

In order to have the necessary strength, paper bags cannot be made entirely of recycled material. This means in part that recycling them has never been widely promoted, and so the paper bag is vulnerable to comparative perception. The upstart plastic bag became the object of very visible recycling programs. In 1990, over ten thousand supermarkets, representing about one third of the total in the United States at the time, were promoting the recycling effort to recover some portion of the almost 400 million pounds of polyethylene then used in the manufacture of plastic bags. As many as one out of every four plastic grocery bags used were returned to some stores, and the recycled material was reportedly made into new grocery bags, trash bags, bottles, and other useful things. But a manufacturer of reusable cloth shopping bags claims that the "average consumer uses 500 disposable paper and plastic bags a year" from grocery stores alone and that "very few plastic bags are actually recycled."

Whether recycled or not, the flimsy, filmy plastic bag, unlike the square-bottomed paper bag, cannot stand by itself and so needs a supporting technology—an infrastructure. At the supermarket checkout counter, supplies of plastic bags are kept at the ready on a wire rack as clever in design as a clothes hanger. It is on this walkerlike frame that the bags are hung in multitudes, opened, and packed in place. As frail as it may at first have seemed, the familiar plastic bag has gained our confidence, convincing us that it is less likely to rip than an overly loaded paper bag lifted by its edges.

When I was a graduate student without a car, I lived in a small apartment that was located almost a mile from the nearest supermarket. The first time I went there, I filled the shopping basket with staples: cans of soup and beans; jars of peanut butter and jelly; boxes of rice and cereal; eggs, bread, and milk. Everything seemed so necessary to get on this first trip that I forgot that at the checkout counter it would all have to be transferred into grocery bags. This was before the plastic bag had made its debut, and so there was no question but that all the groceries I bought would have to be packed into paper sacks or cardboard boxes. I instructed the bagger to pack all the groceries into two sacks, expecting that I would carry one in each arm, with the arm wrapped around the bag's girth. I was warned that the bags would be very heavy, with all the cans and jars in my order, but I insisted. The bags were filled to the brim, and the eggs and milk and bulky boxes of cereal had to be put in a third bag. I figured that I could manage to carry this third, light sack like a lunch bag by grasping its rolled-top edge with the fingers of one of my hands.

This was also before the time when many people thought of appropriating a store's shopping cart and removing it from the premises. It did not even occur to me that I might hijack a cart for the half hour or so it would take me to wheel it to my apartment, remove the groceries, and wheel it back where it belonged. The two stuffed bags were indeed heavy, but I was young and proud, and I wrapped my arms around them as if they were long-lost friends. With the third bag dangling from my right hand, I walked out of the supermarket and headed toward my place. After a few blocks, and out of sight of the clerk who followed me out the door, I had to put the bags down. Luckily, there was a waist-high wall beside the sidewalk, on which I could stand the heavy square-bottomed bags. Had I had to put them on the ground, I might not have been able to pick them up again without ripping the bags or spilling their contents. After another three blocks, I took another break by setting the bags down on the trunk of a parked car. When I reached my apartment, I set the two heavy bags down on the porch and carried the light one inside. It was half an hour later that I went back out to retrieve the others one at a time. Except when I had access to a car, I never purchased such a large order again. On subsequent trips to the supermarket, I seldom used a shopping cart, so that I could buy only what I could carry to the checkout counter, and when I did use a cart, I regularly surveyed what was in it, weighing and packing the order in my mind.

In fact, it was the limits of what shoppers could carry to the checkout counter and home that drove the development of both the shopping cart and the paper shopping bag. Baskets have long been used by shoppers, and in more traditional cultures, people today can still be seen carrying an empty basket to the market and a full one home. A basket carried on the arm can hold only so many groceries, however, and store owners have long looked for ways to increase the amount of goods that shoppers will buy. As late as the early 1950s, when I was growing up in Brooklyn, neighborhood grocery stores relied on clerks to retrieve and bring to the checkout counter what customers stood there and ordered. Boxes and cans were stocked on shelves that reached all the way to the tin ceiling, and the clerks reached the desired item with a long pole fitted with tongs operated by a sort of trigger mechanism. Experienced clerks did not even use the grasping feature of their pole, but merely used the tongs to flick the can or box out from the shelf and caught the falling item with their free hand.

Self-service grocery stores may have been a rarity to me, but they were attractive to owners and customers alike, for fewer clerks had to be hired and selections could be made directly by the shopper—as long as the shelves were within reach. In the 1930s, Sylvan N. Goldman, who was operating a self-service grocery in Oklahoma City, knew that if shoppers could carry more, they would buy more. His first shopping cart looked like a folding chair on wheels, with a folding wire basket sitting in the seat. When not in use, the basket and its supporting frame could both be folded up and stored against a wall, thus conserving floor space in the store. In a subsequent patent, Goldman devised a folding frame that could accommodate two removable baskets, which could be nested for compact storage. The inventive merchant founded the Folding Carrier Basket Company to manufacture and sell his shopping carts. Goldman's folding carriers were soon improved upon, resulting in competing carts with permanently attached baskets. In time, the nesting shopping cart was developed, as was the cart with a child seat. Regular shoppers continue to notice incremental improvements in the shopping cart, which remains far from perfect, as every shopper knows. Indeed, the redesign of the shopping cart was the subject of a *Nightline* television program showcasing the nature of design as practiced by IDEO Product Design, formerly known as David Kelley Design, headquartered in Palo Alto, California. Interestingly, the cart that the project team came up with functioned more like Goldman's removable basket model than those found in supermarkets today.

Shopping carts have not solved the problem of carrying the groceries from store to home, however. As we have seen, this age-old problem led to the development of the paper grocery bag, which has its own limitations, as I learned on my maiden solo shopping trip as a graduate student. But long before my experience, Walter H. Deubner, the operator of a small grocery store in St. Paul, Minnesota, saw his customers struggle and wished they could carry more items out of the store. He devised a shopping bag by reinforcing a paper bag with cord, which also formed handles. The "Deubner Shopping Bag" sold for five cents (a practice long continued by department stores), and by 1915, over a million were being bought each year. Deubner's bags are said to have been "strong enough to carry up to seventy-five pounds worth of groceries." That is a remarkable capacity, but beyond the weight that most shoppers would want to or be able to carry very far. It was no wonder that in 1999, Robert Mentken, a Manhattan inventor, patented a shop-

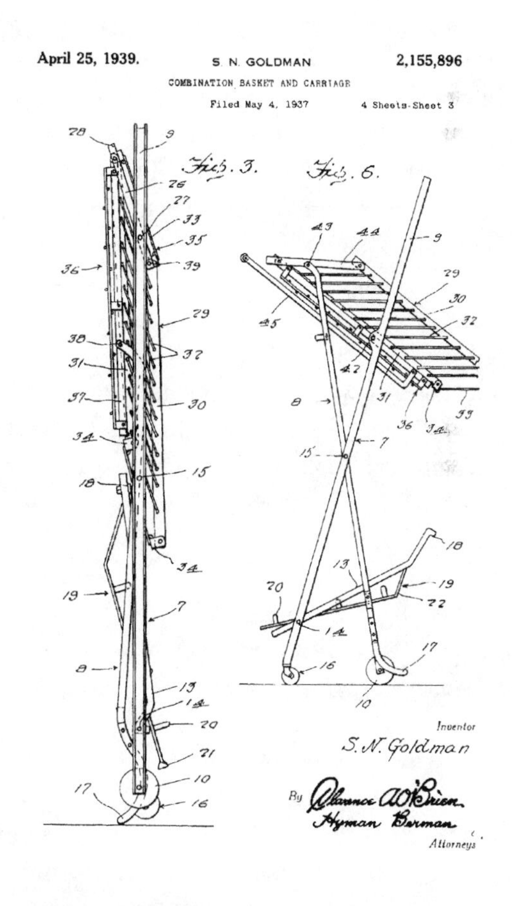

The folding features of Sylvan Goldman's shopping cart were illustrated in his 1939 patent.

ping bag that could be worn like a backpack. He called it the BakSak and expected merchants to like it because it would raise the logos imprinted on their bags to eye level. Long before Mentken's invention, an alternative to a back sack, in the form of a personal shopping cart, was developed for use in urban areas, where wagons, trucks, or automobiles were not driven to the market. This deep, collapsible wire basket on wheels could accommodate at least as much volume, if not weight, as several large grocery bags, but its use has largely fallen out of fashion, except perhaps in places where shoppers still walk to the store. Cars and plastic bags, the descendants of shopping bags, have displaced recycled boxes, brown paper bags, wire baskets on wheels, and just about every other means of conveying groceries from store to home.

The handles of plastic grocery bags, which enable shoppers to carry many more bags at one time than they could if the bags were paper, are among their most competitive features. The paper-bag industry has tried to take away this advantage by introducing handles on its grocery bags, essentially making them into shopping bags not unlike the ones Deubner devised almost a century ago. But, unlike the classic paper shopping bag with its thick string or rolled-paper handles, the new bags have flat handles. These work fine as long as they are pulled parallel to the side of the bag to which they are glued. But they have very little holding strength when pulled away from the bag, such as might happen if a shopper were to try to lift or tote a bag by only one handle. It is no wonder that many of these "handle-bags" carry instructions to "Please hold both handles" and warnings to "Lift with both handles" printed on the bag's bottom. (Interestingly, such bags have a vestigial notch cut into one side of the top edge. It is not very useful for snapping the bag open, however, since the handles shield it from easy access by the bagger's fingers.)

The Paper Bag Council advertises the handle-bag's "strength, durability, and capacity," a clear comparative reference to the plastic grocery bag's apparent flimsiness, its susceptibility to being punctured and split by the sharp corners of boxes, and its smaller capacity. But the cause is likely to be a losing one for the paper-bag industry, given the perceived advantages of the plastic grocery bag for merchant and patron alike.

Though the plastic bag may not yet have been declared perfected, people have come to adjust to its limitations and even to its downright failures.

For all of its advantages when full, there is no way to pack a plastic bag with attention to the same geometrical aesthetic as can be achieved in a square-bottomed paper bag. The best the plastic bagger can do is segregate like products and keep from placing the eggs between the two-liter bottles of soda. Putting a plastic bag full of jars and cans down in a car trunk not designed to keep them in place is an invitation for the contents to roll away and seek out the trunk's most distant corners. The lightly loaded plastic bag containing boxes is likely to end up having pinholes in its bottom, making it unsuitable for saving for reuse. The heavily loaded plastic bag results in an anxiety trip between checkout counter and car. Who has not worried that the bag with the jar of pickles and the special bottle of wine might give out as he steps off the curb? Yet the plastic bag has clearly become the container of choice, shoppers adjusting to its limitations the way people adjust to those of all designs. The once near-perfect upstanding paper grocery bag has mostly been displaced by something that is at the same time superior and yet inferior. That is the way it often is with designed objects.

NINE

Domestic Design

DESIGN IS seeing the potential for a Halloween mask in a paper bag, the makings of a quilt in a chest of remnants, or a good meal in a refrigerator full of leftovers. It is seeing an omelet where others see only eggs and ham and cheese—and a skillet. This is not to say that designing is just like cooking, but cooking a meal or creating a complex dish is designing.

Imagine this scenario. A writer who works in her home in the country has been busy at the computer all day and has not been out of the house. At five o'clock, she decides to drive into town to get some things to cook for dinner. When she goes out to her pickup truck, it will not start, the battery having died. Living in a rural farmhouse some distance from the nearest neighbor, she goes back inside to call AAA to get the vehicle jump-started. The phone, by which she had been connected to the Internet all afternoon, is now dead. The storm that had been threatening must have blown a tree onto the phone line. Looking at the dark computer screen, she sees that the power lines must also have been taken out. The rain is now too heavy to contemplate the four-mile walk to town, and so she decides to wait until the next day to deal with the truck. Perhaps the phone and power will be back by then.

The combination of storm clouds and the last light of day have made the kitchen very dark, but the stranded cybernaut gets a flashlight and lights some candles. Having become accustomed to driving into town late each afternoon as a social antidote to her solitary day at the computer, she has few supplies in the refrigerator or cupboard, preferring, as much as possible, to buy fresh ingredients for the dinners she cooks by herself for herself. This night, she is a terranaut in a rain- and windswept farmhouse rocking on a planet hurtling through space, and she will have to make do with what she has.

She could dramatize her plight by emptying her refrigerator and cupboard and throwing all the supplies onto the kitchen table, but she does not have to do that to know what few things are available to her. She just has to open a couple of cabinet doors and survey the sparsely stocked shelves. But she also has to keep in mind that her electric stove is not functioning now. Still, after working so assiduously all day, she has gotten her mind and body in the mood for a satisfying meal, and she does not want to disappoint them. She will rise to the challenge and make something out of the spare ingredients in her icebox; she is determined to design a meal fit for a queen.

In the mysterious way in which the designing mind works, the stranded chef grasps at once the totality of combinations and permutations that are possible with the supplies on hand and comes up with a menu. She also decides to use as much from her refrigerator and freezer as possible, just in case the power outage lasts so long that the contents spoil. She takes the dozen shrimp out of the freezer to let them thaw while she prepares some accompaniments. She removes the horseradish and catsup from one of the shelves attached to the refrigerator door and mixes up a tangy cocktail sauce. There are some leftover boiled potatoes in a plastic bag on one of the shelves and also some cooked asparagus spears. She uses some mayonnaise and pepper to make a dressing for a potato salad, and some vinegar and oil to dress the asparagus. There is enough in the still-chilled bottle of Chardonnay for a glass of wine. She arranges the candles and some silverware on the table and sits down to the excellent, if cold, meal that she has designed. She does miss her usual fresh cup of coffee afterward, but without that and power to read by, she happily goes to bed earlier than usual. In the absence of the sounds of the electric clock ticking and the refrigerator motor running, she finds herself in a contemplative mood and falls asleep thinking about her day.

Most meals are prepared under more normal circumstances, of course, but the challenge may not be any less. In fact, preparing a hot meal, rather than a cold one, introduces additional constraints and complications, for the various parts of the meal must each be cooked separately and then brought together on the plate while all are at serving temperature. Starting the broccoli or cauliflower too far ahead of the meat's being done means overcooked and mushy vegetables. Miscalculating how long it will take the potatoes to bake can result in a too-well-done steak.

The larger number of dishes and the larger quantities associated with a holiday meal can make the person in the kitchen seem like a chemical engineer designing a process plant. And more often than not at Thanksgiving or Christmas, a couple of ingredients may be found to be missing or in low supply just as they are needed. Time was when supermarkets were closed on holidays, and then the cook was truly left to her own devices. Nowadays, since supermarkets have evolved into hypermarts, many are open twenty-four hours a day seven days a week, holidays included. All it takes to save the project is for someone to make a quick run in and out through the express checkout lane to buy a bottle of gravy browning for the turkey or cinnamon for the baked apples.

It is not only cooking that provides a domestic example of design. Every time we vacuum the house or mow the lawn, we are faced with a design problem. In which corner do we start? Do we proceed in a rectangular or a diagonal pattern? What do we do when we encounter obstacles? Do we move the furniture and spare the wildflowers? Most of us mow the lawn seemingly without thinking, but, in fact, somewhere in the process a design decision is being made, at least the first time we tackle the chore. After that, we can just do it the same way we did it last time. However, there are people who want to improve what they make or do each time they do it, or who prefer to put some creativity and variety into whatever they do just to make it more interesting. Thus, every domestic activity can present a challenge, if it is considered as a fresh design problem.

One of the problems I encounter when vacuuming a long room's carpet is how to make it look freshly done and footprint-free when I'm finished. This means that I should start in a far corner and work my way backward toward the door, as if I were painting the floor. Unfortunately, I may forget that by plugging the vacuum cleaner's cord into a centrally located socket, which allows me to reach all the corners, I have to walk across the newly vacuumed room to unplug the machine. If I am lucky, the cord will be plugged into an outlet from which I can yank it with a firm tug from a distance. More often than not, it will be plugged into one located in such a place that the cord is tugged sideways, therefore bending the prongs. When I do have to walk across the newly raised pile and leave footprints on the otherwise-fresh carpet, my vacuuming efforts seem diminished. By giving some attention at the outset to what we might consider a mindless task, we can both minimize the time it takes to execute it and make the end result more satisfying.

I once gave more thought to mowing the lawn. I would think while walking behind the machine about the pros and cons of progressing in increasing or shrinking concentric rectangles around the house versus working in windrows and completing the grass on one side of the house before advancing to the next area. Our house did not have equal areas of grass on each side, however, and so the first option was not possible. That did not stop me from considering it in the abstract. Since the same amount of grass had to be cut in each case, the former method clearly might have been faster, because no time would have been wasted in stopping to turn the mower 180 degrees at the end of each row. However, employing the unidirectional method would have meant that I was always walking in the same direction, and thus there would have been virtually no variety involved. The act of turning the mower around every now and then would have broken up the monotony and given me a different perspective on the lawn itself. The extra motion involved in turning the mower might have been counterbalanced by a quicker pace, of course, which, in turn, might have yielded the paradoxical result of taking a longer path resulting in a shorter journey. Few design problems have only a single, obvious path to their solution.

It has become customary to mow the grass of baseball fields in textured patterns, which can be very striking for the fans to look at. Achieving the pattern involves using a heavy roller to lay the grass down in the direction of mowing, in order to give it a nap. The practice has not yet become widely used on residential lawns, but a strikingly illustrated story that appeared in the Home section of the *New York Times* not too long ago may have given some homeowners a new perspective on how to design their lawns. Mowing the grass, like any other activity, can be either a chore or an expression of creativity. A lawn designer, even an amateur one, will see a pattern and a path where the toiler will see just grass. Depending on which way the task is approached, midway through the chore, the grass can appear to be half mowed or half unmowed.

Lawns can pose other problems. One big design issue that institutions like schools and colleges face when laying out a new quadrangle is where to put the paths. Because the buildings are often arranged neatly around a rectangle of grass regularly punctuated by trees, the natural thing to do is to locate the walkways around the periphery, thus preserving the symmetry of the cloistered space. (Institutions of free inquiry tend to have curiously rigid architecture.) People do not want to walk around the periphery of a rectangle to get from one side to the other, however, and so designers also often locate walkways crisscrossing the grassy middle. The arrangement of these transverse and diagonal walkways always looks good and right on the plans of the landscape architect. But invariably, once the quadrangle has opened for business, with the classes taught in the buildings changing every hour or so, the students and teachers seek to optimize not the distance between main entrances but those between lecture rooms and offices that may be far from one another. They rush across the lawn to make a beeline from a minor exit of one building to the side entrance of another. Within a very short time, the lawn is worn down, with dirt paths leading from and to where most people are and want to be, regardless of where the official paths are located. Wise designers, especially if they are not privy to the details of how classes will be scheduled or offices assigned in the buildings, at first do not install crisscross paths at all. Rather, they advise the institution to wait until the traffic patterns have been established and then install permanent paths where the grass has been trodden. To lay out the design of a quadrangle with no prior knowledge of its daily use can result in selecting a path less frequently taken, if taken at all. Such decisions can make all the difference in whether the design is a success or a failure.

Some successful designs are a result of dumb luck, something neither foregone with any logic or foreseen with any insight. When I was still a bachelor living with other bachelors, I developed a small repertoire of dishes that I cooked for my roommates whenever it was my turn to prepare dinner. One popular dish of mine was spaghetti with a meat and tomato sauce, the recipe for which I found on a box of pasta. The first time I prepared the sauce, I misread "garlic cloves" as simply "cloves," and so I left out the crushed garlic altogether and added what seemed the right amount from the little can of pungent powder labeled CLOVE that I found on the spice rack. On the basis of nothing at all, I imagined that a single clove was about the size of an acorn, and so I spooned out an amount of powder that I estimated would result from crushing a single clove. Not knowing how wrong I was on several accounts, I mixed the ingredients as instructed on the spaghetti box and let it simmer for the prescribed time. When my roommates came home that evening, they commented on the delicious aroma coming from the kitchen, and I was sure I had a hit. When it came time to serve dinner, I ladled the thick red sauce over generous helpings of pasta and presented it simply as spaghetti and meat sauce. Everyone loved it, and it was welcomed into the regular rotation of dishes that we ate.

One of my roommates was dating a girl who also had a roommate, and so a double date was arranged. Having been smitten, I was grateful to my roommate, and we agreed to double-date frequently. On one occasion, the girls invited us to their apartment for a home-cooked chicken dinner, which we certainly enjoyed. We naturally reciprocated by inviting the girls to our apartment some weeks later. My roommate was to make the salad and dessert, and I was to make the main course—spaghetti with my special meat sauce. The dinner was a great success, from my point of view, and everyone seemed to enjoy the spaghetti sauce. My sweetheart, who really knew how to cook, asked what was in the recipe, and so I retrieved from a kitchen drawer the back panel that I had torn from the pasta box. She read it silently and graciously thanked me for showing it to her. It was only later, after we were married, that she informed me that powdered cloves were not the same as cloves of garlic. Since she loves garlic in her food, she must have known immediately that what I had prepared was missing something expected, but she was too thoughtful to ruin my self-satisfaction with my culinary achievement. Nevertheless, her enjoyment of the dish was certainly sincere, since the clove sauce has become a staple of our own dinners, regardless of who prepares them. I had invented a new sauce, but not by design.

It is not only in the kitchen that such unintended inventions occur. They also happen in the chemistry laboratory, where, in the tradition of science, the phenomenon is not called simply "dumb luck" but goes by the exotic multisyllabic word *serendipity.* One dictionary definition of *serendipity* is "the phenomenon of finding valuable or agreeable things not sought for." The word derives from the Persian fairy tale "The Three Princes of Serendip," and it is often used by scholars and scientists to describe the experience of coming across something by accident while looking for something else. The historian looking through a politician's papers for evidence of a legislative deal and finding a previously unreported sheaf of love poems to his theretofore-unknown mistress would be said to have experienced serendipity. An astronomer looking for a distant star but coming upon a new planet instead might also be said to have benefited from serendipity.

Chemists are always searching for new things, which they call "compounds," hoping that some previously untried rearrangement of known molecules will result in something better than existing arrangements. Because chemists attempt to synthesize new things out of old, they are the scientists most like engineers and inventors. They are not just trying to understand how nature works; they are seeking to design new things out of what nature has provided. Chemistry is also the science most like cooking, for chemists take different ingredients—chemicals—combine them with others, and subject them to, among other processes, different forms of heating and cooling. Chemists use, though often under different names, pressure cookers, ovens, freezers, and whatever else might produce something new that tastes, smells, looks, sounds, or feels good.

In 1938, Roy J. Plunkett was a new research chemist at the Du Pont Company, and his first assignment involved working with Freon, a trademarked substance. Over time, *freon* became the generic term for a variety of nonflammable gaseous and liquid chemical compounds known as fluorinated hydrocarbons or chlorofluorocarbons, a union of elements essential to the operation of refrigerators, air conditioners, and aerosol cans. The fact that freon released into the atmosphere depletes ozone was not known to Plunkett or his contemporaries, and the principal application of the chemical was still in refrigerators. When he went to work for Du Pont, both that company and General Motors manufactured a freon known chemically as tetrafluorodichloroethane, which went by the more simple designation of refrigerant 114. Unfortunately, as effective as it was, refrigerant 114 was not very easy to manufacture, which meant that it was expensive. It was the job of Plunkett and other scientists acting as chemical engineers and designers to find a new refrigerant that could be produced more easily and thus more economically.

Plunkett was trying to make a new refrigerant by reacting tetrafluoroethylene, which is known as TFE, with hydrochloric acid. Since TFE was expensive, he stored it carefully in pressurized metal containers that were kept on dry ice to retain the substance in a liquid state until it was needed in its gaseous form. To maintain an inventory of how much TFE was in each container and to know how much was mixing with the hydrochloric acid, the containers were carefully weighed before, during, and after each use. When it was time to conduct the first experiment, Plunkett's assistant, John Rebok, placed a full canister of TFE on a scale and connected it by tubing to a chamber containing the acid. The valve was opened, but nothing happened. They knew there was no leak in the canister, because it weighed the same as it had when freshly filled. They poked at the valve to see if any debris was lodged in it, but none seemed to be. When the valve was removed entirely, nothing but a few white flakes of unfamiliar material fell out.

Rebok believed that the weight of the canister indicated that no TFE gas had escaped, and so he suggested cutting one of the canisters open to see what had happened inside. When this was done, Plunkett and Rebok saw that the interior of the cylinder was coated with a slippery white powder. Tests run on the substance showed it to have a very high melting point and to be chemically inert—it would not react with anything. Evidently, the combination of pressure and temperature resulting from putting the canisters on dry ice had produced a polymer, polytetrafluoroethylene (PTFE), an unintended new compound with spectacular properties. The substance would be given the simpler name of Teflon, which Du Pont registered as a trademark in 1945.

Teflon, which was very expensive to produce, might have been put on the shelf and forgotten, but since it was chemically inert, it became the focus of military applications. The Manhattan Project needed to process the very toxic and corrosive uranium hexafluoride to produce the isotope needed for the atomic bomb, and the inert quality of Teflon made it ideal for gasket material used in gaseous-diffusion plants. Teflon was also found to be transparent to radar, and so it was used to make the nose cones of proximity-fuze bombs, which can sense their distance from a target. After the war, Du Pont looked to exploit Teflon in commercial products, but since it was still expensive to produce and difficult to work with, it was not believed to be suitable for common consumer products, or so Du Pont thought.

When the Frenchman Marc Gregoire learned of Teflon, he used it on his fishing tackle to keep the line from getting tangled. His wife, seeing how well the material worked, asked her husband to apply it to her pots and pans, which produced equally successful results. The entrepreneurial Gregoire was soon selling cookware coated with what he called Tefal, and eventually he sold over one million such items. One of those pans found its way back to America with someone returning from Paris, and when a friend saw how well the pan worked, he contacted Gregoire in France and suggested that they manufacture the pans in the United States also. The American entrepreneur was Thomas Hardie, a reporter for an international news service, who proceeded to look for a domestic cookware company willing to get involved in the enterprise. None was, and so Hardie imported about three thousand of the French-made pots and pans, which he expected to sell to department stores. The stores did not jump at the opportunity, however, and Hardie could convince only one to place an order. When Macy's put two hundred pans on sale in New York City, they sold out within two days, even though a major snowstorm had slowed traffic on the streets and in the stores.

Soon Hardie had set up a manufacturing plant in the United States; in the meantime, many competitors had begun to market their own coated pots and pans. There was a reason that manufacturers had been reluctant to get involved with the product until a market for it had been established. If Teflon's advantage was that nothing stuck to it, then getting it to stick to something, such as a pot or pan, was no easy feat. Since the material is inert, it could not be chemically bonded to the surface of an aluminum frying pan, and so a mechanical bond had to be developed. One way this could be done was by sandblasting the surface to roughen it, so that the long molecules of Teflon could get a hold in the resulting pits and crevices. Another method was to spray the bare pan with a lumpy ceramic coating, which also gave the Teflon something to anchor to, or with a film of stainless steel, which hardened into a microscopically rough surface. The process was refined by mixing sticky polymers with the nonsticky Teflon and then building up the coating in layers, progressively reducing the sticky content. Mixing some ceramic grit into the final Teflon coat provided an armor of sorts to protect the soft polymer from being cut, scratched, or scraped off.

Applying Teflon to rough surfaces resulted in high spots vulnerable to being scraped bare by forks, spoons, and metal spatulas. Even the more sophisticated buildup of the coating left it vulnerable due to the softness of the material. Hence, special utensils were recommended, if not required, for use with Teflon-coated pans. Despite such precautions, the coating could become soft if heated too much, and so even the softest spatula could ruin it. Generally speaking, no Teflon-coated pan could be guaranteed not to lose some of its coating after repeated use. This meant, of course, that little specks or maybe even larger chunks of the polymer got mixed in with the food that was being prepared and so were ingested. This was not a health problem, however, for the inertness of the material meant that it would pass through the digestive system without being absorbed.

Teflon is a magical material in the kitchen and elsewhere. The story of its discovery shows it to be a classic example of serendipity, and the manufacture and use of Teflon-coated products indicate they are effective examples of compromise in design. The very same nonstick qualities that make Teflon such an attractive material for cookware make it difficult to work with during the manufacturing process. Nothing is perfect, not even the most slippery and most inert material yet discovered—or designed. Design can be both easy and difficult at the same time, but in the end, it is mostly difficult. Teflon may have fallen out of a frozen canister, but getting it to stay stuck to a searing frying pan took a lot of doing. The lesson to designers is, of course: If you can't stand the heat, design a cooler kitchen.

Even the most domestic of maxims and metaphors can reveal a great deal about design and invention. Engineers and scientists have no corner on creativity, and every time we cook dinner, vacuum the rug, or mow the lawn, we have the chance to come up with a new way of doing it. Virtually everything we do presents us with the opportunity to do it better, faster, more easily, or, at least, in a more creative, interesting, and less boring way.

TEN

Folk Design

Everything can always stand a bit of improvement. But if nothing is ever perfect, neither can there be a perfect redesign. This reality does not deter us from trying. It also does not prevent us from fixing what's broke and even fixing what ain't. We would like things to work the way we think they should, but they do not always comply.

Attorney General John Ashcroft, who has been said to see the world in terms of opposites, such as "right and wrong, good and evil, heaven and hell," also sees redemption in a pair of cult consumer products at the opposite ends of the design spectrum. By way of making his point that "the universe is binary," Ashcroft once explained, "There are only two things necessary in life—WD-40 and duct tape. . . . WD-40 for things that don't move and should, and duct tape for things that do move but shouldn't." It is not clear if Ashcroft created his push-me, pull-you witticism out of WD-40 and duct tape themselves, or whether he was merely customizing something in the public domain familiar to Internet surfers. Whatever his inspiration, Ashcroft was stating what many have come to embrace as folk philosophy: "All we need is WD-40 to make things go, and duct tape to make things stop." In other words, with the right attitude and tools, no design problem is too tough a nut to turn or too fast a bullet to catch.

But each of these products is ubiquitous, and they have become metaphors for fixing and jerry-rigging things of all kinds, hardware and software, everything and everywhere. And like all that they fix and cobble together or slip apart, duct tape and WD-40 themselves had to be invented, designed, and developed. Indeed, the story of either one can serve to explain the design of the other and of everything else, for although the things of the world may be classified in binary terms, the process by which they come to be is unitary. The all-pervasive process that we call design may be difficult to define, but we know it when we see it. And we can see it in WD-40 and duct tape.

Leah Beth Ward, a business reporter, once visited the corporate campus of Nike, which is located in a suburb of Portland, Oregon. It is here that the company's footwear is designed, and Ward was tracking down a story for the *New York Times.* In the process of doing so, she observed that in the parklike environment around Nike's campus it was difficult to tell the recreational joggers from the shoe testers and the designers. However, she noticed some runners lugging sketchbooks and other things along with them, and she concluded that "anyone carrying duct tape and scissors is probably a designer at work."

Duct tape has come to symbolize not only the creativity of the working designer running with an idea but also the resourcefulness of the folk designer, the ordinary person faced with extraordinary challenges: The weekend handyman whose hammer handle cracks fishes some of the tough tape out of his toolbox; the hiker whose shoe sole separates from the upper part reaches into her knapsack for some silver tape; the rural homeowner whose mailbox rattles on its post goes into the toolshed to get some tenacious tape; the vacationer whose car's water hose springs a leak opens the trunk to look for some heat-resistant tape; the fisherman whose Styrofoam cooler gets punctured by a pole looks in his tackle box for some waterproof tape; the private pilot who wants, albeit unwisely, to carry some dirt bikes into the wilderness by strapping them onto his wing struts looks in the back of the plane for some strong tape; the chemistry student who tends to knock over her Bunsen burner reaches into a utility drawer to get some handy tape to secure the burner to the lab bench. Most recently, of course, duct tape has been enlisted in the war on terrorism, and it has been stockpiled—along with plastic sheeting—in homes across America.

Duct tape is easy to use, but the name's back-to-back *t* sounds make it hard to say. When asked to repeat the words quickly three times in a row, more than a few people find the term a tongue twister and end up saying "duct tape, duc tape, duck tape," a common mispronunciation. In fact, as if providing an example of some kind of back-formation, such speakers are believed by duct-tape devotees actually to be saying what some claim to be the original name, or at least the nickname, for the stuff. "Duck tape" is also reported variously to have been called "military tape," "gun tape," and "ammo tape" during World War II, when it was invented, developed, and first used. Some informal histories simply refer to it as the "nameless, military-green product from the war." But by whatever name or anonym the tape is known, its story has become embedded in popular culture and has become associated with its own folk traditions.

According to one version, the U.S. military wanted a tape that was "durable, waterproof, very strong, and could be easily ripped into lengths," preferably just by the hands of soldiers on the battlefield. The existing product that seemed to come closest to the specifications was medical bandaging tape, which was made by a division of Johnson & Johnson known as Permacell. A research team led by Permacell's Johnny Denoye and Johnson & Johnson's Bill Gross set to work on the problem and came up with a new tape, one whose strength derived from its cloth core. One side of the cloth was covered with polyethelyne, a resilient and waterproof plastic. The other side was coated with a rubber-based adhesive, which stuck tenaciously to almost everything, even itself. Neither the polyethelyne nor the adhesive contributed much strength, but the cloth provided considerable resistance to the tape being pulled apart either lengthwise or crosswise. At the same time, the cloth-based tape could be easily torn cleanly from the roll by ripping it transversely, the way fabric is ripped from a bolt of yard goods. Though the original cloth used in the new tape may not have been actual duck cloth, duck was used in medical tapes. The new product thus did derive from a category of "duck cloth tape," or "duck tape" for short.

Whatever duct tape was originally and officially called, its creation is a classic example of something designed to order, and the stuff soon became very popular among military personnel. It did shed water like a duck, and it was thus ideal for keeping moisture out of ammunition cases, among other uses. Soldiers used it for makeshift repairs of everything from boots and guns to jeeps and aircraft. It has been said to have been employed to cover up the gun ports on airplane wings so as to reduce drag on takeoff. According to airman folk wisdom, tape that was not pierced after a combat mission identified a malfunctioning gun. Certainly the amount of ammunition not spent would also have provided a clue, but when something is popular and effective, it can get credit for more than it might deserve.

The versatility of the tape made it a natural thing for GIs to bring back to the States. Many of those who did return from World War II with it in their duffel bags found jobs in the housing-construction industry, which took off in the postwar years. Among the uses to which the tape came to be put was to connect together and seal the joints in heating and then air-conditioning ductwork. The original army green (olive drab) color of the tape is said to have been changed to its now-familiar sheet-metal gray, whose metallic cast better matched the galvanized ductwork on which it was used. Thus it began to be called and sold as "duct tape," a name that divorced the product from its military origins.

Even though the new name generally stuck, it is still not nearly as easy to say as "duck tape." In fact, young people who had no idea of the product's World War II roots and applications persistently pronounced its name as "duck tape." This confusion of the tape's name was evidently not lost on one manufacturer, which had its own origins in a small industrial tape supplier founded in 1950 in Cleveland, Ohio, and named after its founder, Melvin A. Anderson. In 1966, a young man named Jack Kahl joined the Anderson company, and in five years he bought it and changed its name to Manco, Inc., thus memorializing Melvin Anderson's initial role. Like many of its competitors, Manco manufactured a variety of tapes, including duct tape. In time, Kahl, apparently a born salesman and marketer, took note of the special place that duct tape had in the hearts and minds of its users and of the fact that many of them pronounced its name as "duck tape." To exploit this happy coincidence, he officially renamed Manco's product Duck tape. He also decided to give customers "a champion who represented the helpful, upbeat spirit of duct tape" by introducing a "friendly and helpful ambassador" named Manco T. Duck to promote it. The marketing ploy was so successful that Manco's entire line of tapes came to be sold under the name Duck tape, including a Duck-brand duct tape with a camouflage pattern marketed as Camo Tape, thus bringing the tape full circle in alluding to its original name and milieu. This is not to say that Duck-brand Camo Tape is designed for military use. Rather, it is considered ideal for hunters and outdoors people, who prefer it for repairing and taping things to their camouflage clothing or equipment.

Jack Kahl was not the only one to recognize the exploitable fondness and devotion of so many users for duct tape. Garrison Keillor has called duct tape "the old reliable" and has had skits about it on *A Prairie Home Companion.* Tim Nyberg and Jim Berg, two other humorists, who bill themselves as the Duct Tape Guys, have published books on the myriad uses for the stuff and operate a Web site on which trivia about duct tape resides and to which fans of duct tape flock. Brothers-in-law Tim and Jim are said to have gotten their start as duct-tape comedians at a 1994 Christmas Eve party, during which a power outage occurred. Jim, a kindergarten teacher, is reported to have said that he "could probably fix the problem with duct tape," and he proceeded to recite other uses for the "ultimate power tool." Tim, a writer and graphic designer who saw promise in the dark, began recording the tips on his laptop, and so a fruitful collaboration was born. The Duct Tape Guys were so successful in bringing attention to the product that, at the change of the millennium, they became spokespersons for Jack Kahl's Duck tape.

Since duct tape does not come with instructions, it doesn't limit creativity, according to the Duct Tape Guys. The many uses of the tape have become the stuff of folktales and handyman legend. Some devotees always carry a roll in their car's glove compartment, in their tackle box, in their luggage, or in their purse. Outdoors, duct tape can patch a fiberglass canoe, a tent, or a tarpaulin. On the highway, it can tie up a loose muffler or a dragging tailpipe. At an accident scene, it can serve as a blood-stanching bandage. In the garage, it can provide a space on which to label paint cans and jars of nuts and bolts. In the yard, it can seal a leaky garden hose or splice a broken rake handle. In the home, it can hold the plaster onto a leaking ceiling or mend a torn shade. On dates, it has been used to fashion an uplift bra for a strapless gown and to seal the split seam on a pair of trousers. None of these quick fixes will render anything new again, but they can make do until a more orthodox repair can be made or a replacement bought.

NASA is said to have a policy requiring that the "gray tape" be carried on every space-shuttle mission, a requirement that may have its roots in the well-known story about the role duct tape played in saving the lives of the lunar-bound *Apollo 13* astronauts. Their rocket took off in 1970, less than a year after men first walked on the moon, but already such missions were considered routine and thus old news for television—until something did not go according to plan. Almost fifty-six hours into the *Apollo 13* mission, one of the spacecraft's oxygen tanks exploded and caused damage that threatened the supply from the backup. In addition, two of three fuel cells were no longer functioning. With the condition of their life-support system and their power supply rapidly deteriorating, the crew shut down the crippled command module and retreated into the lunar-landing module of their spacecraft. In the meantime, Mission Control in Houston tried to figure out how to save three astronauts who were 200,000 miles from earth and hurtling toward the moon.

In *Apollo 13,* the motion picture retelling the story of the mission, perhaps the most dramatic moment occurs when engineers realize that unless something is done soon, carbon dioxide will build up to dangerous levels in the makeshift lifeboat. During the actual mission, the astronauts had to continue on their trajectory so that they could make use of the moon's gravity to redirect them back toward earth, but that maneuver would keep them in the confined quarters for longer than anticipated. The lunar module had been designed to support two men for two days, rather than three men for four days, which is what it was now being asked to do.

The astronauts did have on board a supply of lithium hydroxide canisters, which would normally have been used to filter the carbon dioxide out of the air in the command module. However, those same canisters could not readily work in the lunar module because the receptacles for its filters were round and the command module's were square. In what is, for engineers at least, the most memorable scene in the movie, a box full of assorted parts and supplies is emptied onto a table at Mission Control and the engineers are told that these are the only things that the astronauts have on board to work with in order to fit a square peg into a round hole. The ground-based engineers work frantically for a day and a half to jerry-rig something to fix the broken carbon dioxide filter system and just in time come up with a scheme using plastic bags, cardboard, and duct tape.

The fix worked, and though their lunar landing had long since been aborted, the astronauts continued on their journey toward the moon. When they were close enough on the return home, they changed back to the command module and powered it up. It was by no means a foregone conclusion that the craft would still function. In the cold of space, moisture had condensed everywhere inside the vehicle and there was fear of a short circuit. The system did respond, however, and the astronauts were able to land safely on earth, albeit in a capsule in which it rained as the heat of reentry caused precipitation in the moist cabin. The mission has been described as a "successful failure," because the emergency ad hoc engineering action worked. The engineers in Houston designed a way for the astronauts in the *Apollo* space capsule to fabricate and install a filter that was not pretty but that saved their lives.

Years after *Apollo 13,* when the more commodious space shuttle was the focus of NASA's attention, the septuagenarian astronaut-cum-senator John Glenn became the oldest man to travel in space. When asked after a speech what everyday products were used on such missions, he mentioned "Velcro and plain, old, grey duct tape." As the crowd laughed, he added, "I don't know if we could conduct a space mission without duct tape," then explained that it enabled astronauts to post mission notes and to prevent a meal from floating away in the weightless environment.

Sometimes the quick fix becomes the permanent fix, of course, and duct tape memorializes many an averted crisis. The existence of something like duct tape may also inhibit innovation in design, however. The audiovisual technician setting up microphones for a meeting often has a roll of duct tape hanging from his belt, ready to dispense silver-gray foul lines over the handsome carpet of an otherwise-formal auditorium or dining room. The tape molds itself to the wires along the floor and anchors them in place. No one trips taking a seat, and no mike cord is pulled loose during a speech, but the space around the speaker can resemble a basement or a garage workshop. The use of the tape in such circumstances is a convenience for technicians, and therefore they are not likely to complain about the poor design of dining rooms and auditoriums for the purposes to which they are so often put.

There is also a darker side to the use of duct tape. Like any product of design, the stuff can be employed for ill as well as for good. Duct tape was used in assembling pipe bombs that were placed in rural mailboxes in several midwestern states over a few days in the spring of 2002. But some unexploded bombs found in Nebraska mailboxes were not rigged the same way as those placed in Illinois and Iowa, and the difference was believed to be because the bomber ran out of duct tape. Without the tape to hold the trigger mechanism in place in the mailboxes, the bombs were not effective. Later that same year, a kidnapped Philadelphia girl was bound with duct tape and left alone in the basement of an abandoned house. Fortunately, she designed her own escape by gnawing through her bonds.

For all of its putative indispensability, duct tape does have its shortcomings. No matter how beautiful its faithful users may view the process of applying duct tape to the world as being, most applications of the stuff are as downright ugly as the audiovisual technician's floor stripes. The tape is, of course, notoriously difficult to use, because it tends to stick so readily and tenaciously to itself during the process of peeling it off the roll and applying it to things intended and unintended alike. In spite of its name, inferior duct tape is not sufficiently heat-tolerant to withstand the temperatures in furnace ductwork, and it does not perform well in any extreme conditions of heat or cold.

(The term *duct tape* cannot be used in California unless the stuff meets "certain heat-resistant standards.") Duct-taped items tend to be lumpy, messy, and dirty looking, full of loose ends of tape that stick to dirt before they can stick to one another. The perfect tape is, in short, decidedly far from perfect in its use.

Since fresh duct tape does stick so well, its greatest fault may be the difficulty of removing it and its residue completely from a surface to which it has been applied. Perhaps not surprisingly, given Ashcroft's maxim, one recommended way to remove duct tape is to apply WD-40 to it. Like many another useful item, WD-40 had its origins in war and in the military-industrial complex. During the early years of the Cold War, a big problem existed in corrosion attacking the outer skin of aircraft and missiles. One fledgling firm, Rocket Chemical Company, with a staff of three working out of a small laboratory in San Diego, California, set itself the problem of developing a product line of rust-prevention solvents and degreasers that could be used in the aerospace industry.

Rocket Chemical, founded by chemist Norm Larsen, believed that one way to prevent rust was to develop a substance that would keep water, the root cause of the corrosion, from coming in contact with a metal surface in the first place. Thus, the researchers were looking for a "water displacement agent," and to distinguish their various attempts at coming up with a successful chemical concoction, they had numbered sequentially the water-displacement formulas they tried as WD-1, WD-2, WD-3, et cetera. Evidently, all designs from WD-1 through WD-39 had failed to perform satisfactorily, although they had provided valuable information. Each WD experiment had taught Rocket Chemical, as all of Thomas Edison's failed experiments when searching for the proper materials for a lamp filament had taught him, what did not work. In 1953, on their fortieth try, Larsen and his colleagues formulated something different, and it worked. The result was WD-40.

This much of the story seems sound, but the folk history on the World Wide Web is full of embellishments that are more difficult to evaluate. The Duct Tape Guys, who have widened their interests to include WD-40, credit Larsen's wife with discovering household uses for the aerospace product. In addition to silencing squeaky hinges and loosening stubborn bolts, it is said to be useful for cleaning, waxing, and polishing furniture. In any case, in 1957, WD-40 began to be offered in spray cans and sold outside the aerospace industry. In 1969, Rocket Chemical Company changed its name to the WD-40 Company, after its only product.

But the water-displacing solvent in the "can with 1,000 uses" still does have some sticking points. Though it has been used to clean automobile windshields, it leaves streaks. The back of the spray can comes with a list of cautions headed DANGER, including the warnings that its contents are flammable, under pressure, and harmful or fatal if swallowed. Perhaps the product's most damning operational fault for a long time was what to do with the thin red tube that comes with the blue-and-yellow spray can so that the water-displacement agent can be applied with some discrimination to hard-to-reach places. The tube had a propensity to be nowhere to be found when needed most. I used to put a rubber band around the can to keep the tube with it, but before I had use for a whole can of WD-40, the rubber often deteriorated and the tube fell who knows where. The tube can be duct-taped to the can, of course, but then there is the risk that it will not come free without being damaged.

The manufacturers of WD-40 responded to the problem of the errant tube by introducing a pair of notches into the top of the spray can's red cap. The tube, which comes Scotch-taped to the side of each fresh spray can of the degreaser, once removed, may be pressed into the top's notches for storage. However, so stored, the four-inch-long tube overhangs the top and the edge of the two-and-a-half-inch-diameter can and is thus susceptible to being pried loose from the notches, snugly as they do hold on to it. A shorter tube would not be as useful, and a larger-diameter spray can would not be as efficient in containing the internal pressure. Even the most useful of things are impossible to design without compromise.

But there appear to be no limits to the imagination and creativity of the cult followings that duct tape and WD-40 have developed. As the

The tube supplied with a spray can of WD-40 can be stored on the notched cap.

Duct Tape Guys have documented on their Web site and in their books, there seems to be no end to the uses that ordinary folk have devised and designed for these endearing products. They have used WD-40 to take ink stains out of blue jeans, remove old cellophane tape, clean bedpans, dissolve glue, soften leather, repel pigeons, kill insects, keep grass from sticking to lawn mowers, and get peanut butter out of dog hair.

For their ardent fans, duct tape and WD-40 may be the yin and yang of all things, but there is more to design than just stop and go. Many of the greatest challenges to designers lie in problems of degree and flexibility. Just as one size cannot ever really fit all, so no thing need always move or always stay put. Design must deal with the transitions between hold and release as readily as with the extremes. The way to accomplish this with style is not with duct tape and WD-40 but with specific grace within specific constraint, and with brilliant choice leading to elegant compromise. Adhesives and lubricants alone are not enough to save bad design.

ELEVEN

Kitchen-Sink Design

One of the most successful postwar products had its origins in a 1930s incident that involved an unpleasant surprise with a water faucet. At the time, sinks in work areas like basements and garages, and elsewhere, could often be found fitted with two separate spigots located a good foot or so from each other. A rubber stopper or old rag was used to plug up the drain so that the water level rose as the hot and cold streams mixed to the right temperature in the deep tub before being used for washing clothes, mops, and dogs. In 1937, such a sink was being used by Alfred M. Moen, then a mechanical engineering student at the University of Washington, who was working in a Seattle garage to help pay for college. While washing up after work one day, he scalded his hand with some unexpectedly hot water. The incident made him determined to develop a faucet that would prevent such a thing from happening.

There had been earlier attempts to make water faucets more user-friendly. Instead of having two entirely separate spigots, each of which provided water that could be either too hot or too cold for comfort, some plumbing fixtures fed the hot and cold water into a common outlet, where the extremes could be mixed to the desired temperature. But such fixtures did not eliminate completely the problem that Moen had experienced. Turning on the hot water seldom results in hot water right away, for in the time since the last use, the water in the pipe between the water heater and the sink has usually dropped close to room temperature, often to that of a cool basement. If the sink is in an unfamiliar house, workplace, or hotel, it is difficult to know how far the hot water has to come from the water heater or how high the temperature on the water heater is set. A cautious person might hold one hand in the stream and the other on the cold-water faucet, ready to add cold as the hot water begins to arrive. But not everyone always remembers or cares to approach such a seemingly simple system as a sink and a faucet (or two) with such caution and attention.

Even today, and even when using sinks with which I am familiar, I will on occasion experience uncomfortably hot water while washing my hands. The first-floor powder room in my home is located directly over the water heater in the basement, and so the water in the pipe is always a little warm and gets hot very quickly. Still, I often think that by the time it gets really hot, I will have finished washing my hands and will not need to have wasted any cold water. The kitchen sink, on the other hand, is some distance from the water heater, and so the water runs cold for a good while before getting hot. In our summer house, the locations of the kitchen and bathroom sinks are reversed relative to the water heater, and it is the kitchen sink that provides hot water almost on demand. As few and simple as the combinations are, now and then I find myself forgetting where I am and absentmindedly put my hand in an uncomfortably hot stream of water.

The ideal way to use a washroom sink, especially one with separate taps for hot and cold water, is, of course, the same as the proper way to use a washtub: by putting a stopper in the drain and mixing the water in the basin to just the right temperature for washing. Most washbasins are fitted with convenient mechanical stoppers, but filling the basin seems to be done less and less often, perhaps in part to conserve water. (Taking a quick shower consumes less water than having a bath, we are reminded often during times of short supply.) Sinks in public washrooms are seldom fitted with stoppers, perhaps for sanitary purposes or perhaps to minimize the chance of someone leaving the water running and thus causing it to overflow onto the floor. But whether for reasons of conservation or fear of litigation, the water in public rest rooms seldom gets too hot—if it can get hot at all. In fact, with the automatic shut-off features of most such sinks, sometimes the water stops flowing in the time it takes the hand to move from the handle to the stream. Faucets fitted with sensing devices that turn the water on as the user approaches have helped somewhat, when they do not appear to be playing with us, but these latest are inventions that followed by almost half a century Alfred Moen's experience in the Seattle garage.

Moen's invention was, of course, the now-familiar faucet with a single lever, which works somewhat like a joystick, or a single pull-push knob used to regulate at the same time the amount of flow and the temperature of the water. In some cases, the faucet-control device can be oriented to a preferred mix of hot and cold, so that at equilibrium the stream of water is at the desired temperature. In the late 1930s, Moen single-handedly built prototypes of his faucet and showed them to potential backers and manufacturers. Their initial reaction was one of skepticism: The kid's redesign of the faucet pair was too radically different. As inconvenient and potentially harmful, due to scalding, as the existing system was, most people had become familiar with it, and their hands had become accustomed to turning two separate handles for hot and cold water. It was the most natural thing to use two hands to turn two handles, even if not always in the same direction. (Unfortunately, the direction in which hot and cold faucet handles must be turned to regulate the flow of water all too often seems to have been left to the whims of the designer, manufacturer, plumber, or handyman. Their nonstandardized operation is especially apparent to those who travel a lot and stay in many different hotels.)

The idea of equipping a sink with a revolutionary fixture capable of single-handed control proved not to be an easy sell. Still, Moen continued to improve his design and refine his prototypes, but in the meantime, materials became scarce. Metals for civilian use were in especially short supply during World War II, and so Moen's faucet could not be manufactured, even if some company had wanted to make it. He found work as a shipyard tool designer and served in the navy. It was not until after the war, and a decade after Moen invented his faucet, that a Seattle metal-products company agreed to manufacture the clever but curious new device. Only 250 were sold in the first year, but the postwar

This shower-control device, whose single knob can be rotated left and right to regulate hot and cold water and can be pulled out and pushed in to regulate flow rate, is a descendant of Alfred M. Moen's 1930s invention.

building boom provided a ready market for the ingenious plumbing fixture. In time, of course, Moen-type faucets became widely accepted in America, and by the end of the century, they were being installed in new homes more frequently than not. But not in Britain, where as late as the beginning of the twenty-first century separate hot and cold "pillar taps" continued to dominate sales. Even in the world marketplace, cultural forces still exert great influences on design and use.

This is not to say that Moen, whose business card identified him simply as "inventor," solved all the problems associated with hot water in the United States or elsewhere. As clever and convenient as the single-handled fixture is, it does not eliminate completely the danger of scalding water coming out of the tap. The same problem that the young Moen experienced while washing up after work can still be encountered when using his faucet or one descended from it. Indeed, the problem has more to do with the kind of system being used than with the type of fixture, and this fact was masked by the novelty of the Moen faucet and its clear advantages over the old designs.

Just as the water that flowed out of a lead pipe into the pool in a Roman villa was but the end result of a long journey along an aqueduct that ran through the hills surrounding the city, so the regulation of water from a modern faucet is but a drop in the ocean of technology into which it is tapped. Such systems are simultaneously large and robust and small and fragile—from the ultimate source of water in rain, which feeds streams, rivers, lakes, aquifers, wells, and reservoirs behind dams; through water-treatment plants and water pumps; through channels and tunnels and mains and pipes; into water tanks and water heaters and watercoolers; and through the faucet or the spigot or the nozzle into pots, pails, pitchers, glasses, and over hands; and then into the sink and down the drain; and finally to wastewater pipes and then treatment plants, where the water is returned to the supply chain. Just think what happens when any one of the components in the great network fails.

A system of public water supply involves not just the hardware that catches, contains, chlorinates, carries, and conducts the water to our faucets; it also includes the laws and regulations that politicians write, under the influence of lobbyists, environmentalists, farmers, and recreationists. Great compromises of legislation, regulation, and public policy govern how many parts per billion of this or that microorganism or metal ion and what degree of turbidity reaches our faucets. And the system does not end there. Every faucet is touched by the hand of an ultimate user, who trusts in the entire system and each of its components, confident that what comes out is not poisoned, contaminated, caustic, or otherwise unsafe or unpotable. The ultimate user also counts on the system, albeit a more local and personal one, to ensure that the water is not too hot.

At first, Moen focused his design only on the mechanics and

ergonomics of delivering water that would not scald the hand. He succeeded in those aspects admirably, but he also failed to solve once and for all the problem that had motivated him to invent a new faucet. Indeed, the very success of Moen's design has led to a failure of a larger system, for in trying to distinguish their competing designs, designers have left no standard as to how faucets are operated. As anyone who has used a single-handled or single-knobbed faucet in an unfamiliar sink or especially in a strange shower knows, it is still all too easy to get scalded. It is not only the mechanics of turning left or right, clockwise or counterclockwise, or pushing or pulling that matters. If we do not know how far the water heater is from the faucet, if we do not know how high the setting is on the water heater, if we do not know where to set the handle to get a nice compromise between hot and cold, and if we do not know how long to wait before jumping into the shower, then we are no better off than a student at a garage sink fitted with two separate water taps.

This is not to minimize the brilliance of Moen's redesign of the water faucet, but throughout the half century of distraction over its seemingly revolutionary mode of operation, it has been connected to pretty much the same system of water supply and demand that existed in Seattle in 1937. The ultimate operators of that system include human beings in strange hotels with shower controls that can defy decryption. The unknown plumbing can, in some cases, even have the hot and cold water supply crossed behind the marble of the sink or the tile of the shower. Perhaps this was done inadvertently by a plumber who worked on too many rooms that day, or perhaps it was done deliberately by a disgruntled maintenance worker who wanted to get even with the hotel chain by inconveniencing its guests. Who knows what technological mutants lurk behind the polished tile?

We should not become paranoid about any technology, but we should use all technology with a cautious skepticism. A water faucet, like any other interface between us and a larger system, has features and faults. The features are the things the manufacturer's marketing and sales staffs trumpet; the faults are the things they may not even know about. No made thing is without features and faults, and most things have so many more good features than annoying faults that we should not throw out the newborn faucet with the old bathwater. We can embrace the good and outsmart the bad aspects of technology. We can maximize the benefits and minimize the liabilities. We can approach all technology with confidence and caution. There may be the risk that scalding water will come out of any faucet, especially an unfamiliar one, but we still need to wash our hands. We must realize that each new faucet is attached to a local system that may or may not have local quirks, and we can test the system with a cautious finger the way a mother tests her baby's bathwater with a crooked elbow.

Every design, redesign, and resulting piece of technology is a compromise between features and faults, between good and bad, between hot and cold. When engaged in design or redesign, whether as inventors or users, we can adjust the sides of our brains the way we adjust the faucets on our sinks, taking care not to be overly left or overly right, overly slow or overly fast, overly hard or overly soft, overly hot or overly cold. And what the correct compromise is depends upon such variables as the time of the day, the season of the year, and what we want the technology to do for us at any given time. A kitchen knife is a pretty basic piece of technology that can be used to prepare dinner or to commit a crime. But no one claims that the chef's knife should be redesigned to eliminate its pointed tip. Should we want to give up the civilizing effects of fine dining for fear of the occasional fight in the kitchen? Or should we redesign the environment to keep the heat of battle out of the kitchen?

Under no circumstances can we fully escape risk, because we can never escape design. Virtually everything we do involves design in one form or another. In washing up before dinner, we design the temperature and the rate of flow of the water in the basin. We control the technology of the sink, but, since we do this daily, we come to do it unconsciously and automatically. We may not even realize that we have custom-designed a procedure: from the temperature and amount of water we draw from the faucet, to the order in which we soap our hands, to the method we use to wash and rinse, to the way we dry our hands and face, to the way we return the towel to the rack. But each of us has, through trial and error and practice, designed a process that we might even call a minor ritual. It is as much a design as Alfred Moen's faucet, and it enables us to use such a device without having to think about it.

Design so pervades our everyday activity that it is something we cease to see as design. We learn from childhood how to pick and choose among all the contradictory ways that our parents, relatives, friends, and teachers have shown us how to do things, from washing our hands to thinking our thoughts. We rebel against, embrace, adopt, adapt, and compromise among the possible choices in order to design our own way. Sometimes the differences are barely perceptible, but we all do wash our hands differently. We each like our water at a different temperature. But sometimes when we do get even a Moen faucet running at just the right temperature, someone downstairs flushes a toilet and adjusts another Moen faucet to just the right temperature to wash his hands, and the water temperature we had so carefully selected before stepping into the shower suddenly rises and then falls. As we each adjust and readjust our faucets, we affect the other's. As perfectly as each of the faucets might work alone, they seem to work chaotically when used at the same time in a larger system and context. Such are the frustrations of design and the use of designs.

Design and redesign involve everything and the kitchen sink, but this is not to say that redesign always results in an unqualified improvement. The mousetrap is a cliché of design, of course, and part of the cliché is that consumers are ever poised to beat a path to a successful inventor's door. The same cannot be said of some equally mundane household products, even though these products look as ungraceful or work as poorly as the standard wood and wire mousetrap. One such product is the familiar and all but invisible tool once found in virtually every kitchen drawer. Until a decade or so ago, the clever design of the generic potato (or vegetable) peeler had remained essentially unchanged for over a century. The all stainless-steel model made by Ekco, which is sold as a "floating blade peeler," may be taken as representative: It has an open handle made of a strip of steel bent and welded into an elongated tuliplike shape. But the handle has narrow edges that dig into the hand and a short body that is uncomfortable to hold. The tool's loose-fitting blade is attached to a long nail-like shaft that is exposed and rattles within the handle, small extensions of which keep the blade from swiveling too far one way or the other from the job at hand. For some users, the tool is annoying to listen to, not to mention prone to mistake the skin of a bare knuckle or a fingertip for that of a potato or a carrot. In its more modest manifestations, the peeler can collect gunk in its crevices and, when not stainless steel, rust on its handle. But for all of the minor faults of this familiar kitchen tool, until recently the culinary crowd did not seem to be clamoring for a new, improved design. After all, the thing worked.

However, in the late 1980s, the faults of the potato peeler came under the scrutiny of Sam Farber, who by then had had decades of experience with products for the home. Sam was a nephew of Simon W. Farber, a Russian immigrant tinsmith who established the S. W. Farber Company in 1900. Ten years after starting out making copper and brass bowls and vases, the company began making the Farberware line of kitchenware, for which it became well known. After World War II, when the materials became available for domestic products, Farber introduced stainless-steel and aluminum cookware. Sam's father, Louis Farber, owned Sheffield Silver, where young Sam worked for eleven years after his graduation from Harvard with a degree in economics. In 1960, Sam Farber stared his own company, Copco, which became a maker of "sleek-looking steel and enamel cookware" perhaps "best known for its colorful cast-iron and enamel tea kettles with teak handles." He sold the company in 1982 and retired to the south of France, where he expected to collect art and write. But he soon found that he "missed the social contact of work," and so he began "cooking up another business idea."

His wife had developed arthritis and so found traditional food-preparation utensils like the potato peeler uncomfortable, if not painful, to use. There were alternatives to the old peeler on the market, but their bulky handles were said to make them look ugly, and there was often a perceived stigma to their use. Farber saw the opportunity for a new product line of kitchen gadgets that would be both user-friendly and aesthetically innovative, so as not to stigmatize the user as handicapped. In the terminology of some students of product development, he perceived a "product opportunity gap," which could be filled with clever design and marketing. To test his hypothesis, Farber talked with those who bought merchandise for stores. According to Farber:

> I heard a lot about better packaging and displays, assortments that were too large, the need for larger retailer margins, but nothing about the failings of the products. . . . I asked what faults [the merchandise buyers] found with products on the market and received answers like "Some are good and some are bad" or "They've always been like that."

Farber was convinced that he had "a winning idea." He also, if perhaps fortuitously, anticipated at the beginning of the prosperous 1990s that it would be a propitious time to introduce a new potato peeler and other easy-to-hold kitchen utensils, even if they would cost many times more than the serviceable items they were designed to replace.

To flesh out the concept and get down to details, Farber involved the New York firm of Smart Design, persuading it to accept a small advance and royalty agreement in lieu of its usual fee. In this way, Farber reasoned, he not only would reduce his up-front costs of starting a new business but would also get the designers to view the product as an investment of their own. The problem given to Smart Design was "to come up with tools that were comfortable in the hand, dishwasher safe, high quality, good looking and affordable." (Farber "didn't want a $20 peeler.") Though initially motivated to put comfortable kitchen utensils into his wife's arthritic hands, Farber had broadened his goal to a much wider market: "Why shouldn't everyone who cooks have comfortable tools?" Smart Design asked questions of a different kind and began to do research of its own:

> The design team talked to consumers, examined and used competitive products, interviewed chefs, and spent hours with volunteers from a New York arthritis group to learn the problems of hand

movement. They delved extensively into the range of manual limitations, from serious permanent disabilities to the limited mobility and declining strength associated with aging. They also noted gadgets with rusting metal and cracking plastic, dull peeler blades and can openers that didn't cut. Like Farber, their passion and belief in the project grew.

After much research, analysis, designing, and testing, a new product emerged. Though the slotted, double-edged stainless-steel blade of the resulting "swivel peeler" was to be pretty much the same as its nineteenth-century predecessor, it was to be supported and shielded by a graceful arc of hard black plastic, whose extension would serve as a tang to connect the blade to the handle. The handle itself is the part of the new tool that became its cachet. The shape of the ample handle is oval in cross section, presumably making it easy to grip, comfortable to hold, and unlikely to turn in the hand. The relatively massive handle might have given the new peeler the same awkward look and stigma that marked earlier attempts at kitchen utensil redesign had it not had some further innovative features. These included the use of a new material, one that would be soft enough to deform somewhat under a user's grip but at the same time would remain stiff enough to hold its overall shape. Furthermore, the material had to have a surface sufficiently rough to keep the handle from slipping in a wet hand. The handle was also to be molded with flexible fins on either side—visual signals of how to hold the tool—that would yield under the thumb and index finger to provide an even better grip. It was also to be made with a large tapered hole near the end, ostensibly to make it easy for a shaky hand to hang the peeler on a hook, but also to save a bit of material and to break up the massiveness of the handle itself.

The handle material that the product developers chose to use is known as Santoprene, a processed rubber (technically, a thermoplastic elastomer) used for dishwasher gaskets. However, as is so often the case in design, some of the desirable features proved to be incompatible with others, at least according to the conventional wisdom of the members of the manufacturing community who were first approached to

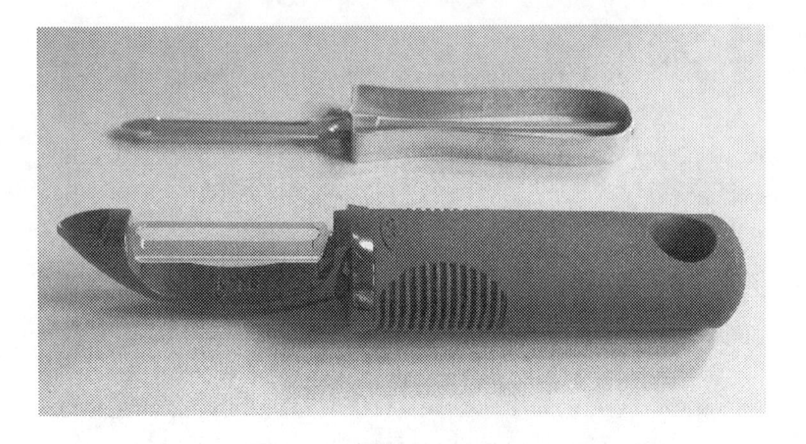

Stainless-steel Ekco and Santoprene-handled Oxo Good Grips vegetable peelers each have their design adherents and detractors.

make the handle. These manufacturers said that the relatively delicate fins, termed "fingerprint softspots" by the designers, could not be molded in Santoprene because the material was not strong enough to be used for such slender structures. Since the flexible fins were an important feature ergonomically and aesthetically and gave the peeler a desired quality—and a designed look—the developers did not wish to alter the appearance of the fins by making them more structurally stout. The persistence of the designers paid off, for eventually a manufacturer was found in Japan who demonstrated that the shape could be molded with fins as designed, and the "softspots" became a virtual trademark of the handle.

In 1989, Farber established the firm Oxo International to make and market the new utensils. While looking like a line from a game of tic-tac-toe, the unusual name of the firm "doesn't stand for anything," but Farber liked the word "because it read the same rightside up and upside down," especially when printed in all capitals—OXO—as it usually is on the company's products and packaging. The name itself is thus a brilliant design, one that is hard to misread or overlook in retail displays.

The new peeler and companion utensils were marketed under the brand name Oxo Good Grips, and the handle design in particular proved to be very successful. In fact, to capitalize on that success, the company emphasized the grip—especially the fin contours, which echoed and were echoed by the curve of the peeler's plastic blade guard—on many other products (which grew in number to about three hundred). Oxo Good Grips kitchen tools "won both customer approval and critical acclaim, garnering almost every major design prize."

The case of the potato peeler and other Oxo utensils is representative of the design subtlety and complexity of even the simplest of things. There is hardly an artifact imaginable that does not have competing features, conflicting qualities, or clashing requirements to test the resolve and mettle of engineers, industrial designers, and manufacturers alike. Even when the challenges of design and manufacture are overcome admirably, as in the case of the Oxo swivel peeler, there is rarely unqualified success. As handsome as a product might be, as well as it might work, and as gracefully as it might wear, there is always some new challenge to be met. That challenge often comes in the form of some economic or competitive issue.

When it became too costly to continue with the Japanese manufacturer who had developed the processing standards that made the Oxo peeler's handle design workable, a new manufacturer in Taiwan was engaged. Such change invariably introduces subtle variations in a successful design and the perception of it, even if nothing more than that "Made in Taiwan" replaces "Made in Japan" on the packaging. Furthermore, with the success of Oxo products, imitators and innovators have followed suit, creating competing kitchen utensils, each playing in some implicit or explicit way on the limitations of the leader, which can include price. (In anticipation of this, Farber and his colleagues had decided on a strategy to "knock off their own product before a competitor did," and so they began introducing lower-priced alternatives.) As "perfect" a solution to a design problem that an innovative product might appear to be, there is always some new competitor to fend off, something latent in the design to criticize.

The Oxo peeler is a striking design, and its innovative handle was rightfully the focus of attention when it was introduced. But as any product becomes more familiar, it comes to be viewed as less remarkable and is seen in a less idealized light. When the peeler is not in the kitchen drawer (who really hangs such a gadget on a hook?), it is in the wet hand of someone peeling potatoes or carrots. Under such conditions, anyone's thoughts are less likely to be about the utensil's wonderful grip, which by now has come to be expected of such utensils, than about some other aspect of food or about the joys of eating in restaurants. Even the arthritic cook is less likely to pay attention to the easily gripped handle than to the busy blade. When this happens, the gracefully arcing plastic guard can become less appreciated for its aesthetic lines than despised as an obstacle to seeing what one is doing to a vegetable. The plastic shield does not look like so beautiful a design feature when it blocks the eye of a potato from the eye of the beholder. Furthermore, like a rusting old plain steel peeler, the new one's blade in time has to dull. While this may be ever so slight at first, it can produce just enough added resistance on the potato skin to call attention to the deterioration of the blade relative to the stout handle's survival. After all, the ultimate purpose of the tool is not to be held but to skin vegetables. So why is the blade so small relative to the handle? When looked at in comparison to the now-familiar thick handle, the business end of the tool can begin to seem puny. If that observation is made, the peeler sitting on the counter or in the sink can appear malproportioned. And once that observation has been made, the design is vulnerable on still other counts. The next time it is picked up, it might not feel as comfortable in the hand as it once did.

This should not be surprising, for the design of the potato peeler, like that of everything made, involved compromise. As much as the old standard peeler's handle and blade support were redesigned, the cutting blade stayed pretty much the same. (That essential feature of the classic design is not easy to improve upon or desirable to change, though one user of an imitator peeler found its blade cut too deeply into vegetables. He flattened the shiny new blade with a pair of pliers.) But even in the radically redesigned handle of the Oxo Good Grips, there had to be compromises between fitting the small bony hand and the large hammy one. Such a compromise is usually disguised under the rubric of "fitting the fiftieth percentile person," but in the 1990s, would that statistical person have been a man or a woman? Given that the handle had to have a specific size, who would have had the nerve to think that something that would be held potentially in billions of hands could fit into each of them perfectly? Certainly there are some for which it is too fat, some for which it is too slender. For me, the handle's girth is good, but the length is short. The one-size-fits-all dimensions of the handle of the swivel peeler were necessarily a statistical compromise. Could it be any other way? Is it any other way with any other one-size-fits-all thing?

TWELVE

Off-the-Shelf Design

ALL DESIGN PROBLEMS are difficult, if we are to judge their solutions by the strict but fair standards of functionality, ergonomics, aesthetics, manufacturability, desirability, durability, maintainability, acceptability, affordability, and countless other criteria relevant to the specific design choices at hand. If such a seemingly simple problem as designing a good potato peeler is fraught with obstacles to achieving unqualified success, then how much more difficult must it be to produce a better mousetrap or a best chair?

The basic design requirements for general-purpose seating, say in a restaurant, are relatively simple: Support a diner at a table for the duration of a meal. Implicit in such requirements are the real challenges: Support a diner who may be a two-year-old child and can walk upright under the table (no wonder there are booster seats and telephone books) or a basketball player who must duck to pass though the doorway; support the diner so that he or she can sit at the right height and close enough to the table so that the hands and arms can move as necessary; support the diner comfortably for the entire meal, which may last ten minutes in a fast-food establishment or three hours in a fine French restaurant; support the diner in a style appropriate to the prices on the menu. To these may be added some ancillary requirements relating to how the chair can be moved to facilitate sitting down at and getting up from the table; how the chair can be stacked to turn the dining room into a reception room; how the chair can be cleaned after a careless child, or adult, spills a drink on its seat. Charles Mount, who designed nearly three hundred restaurants, believed that a chair in a fast-food establishment had to satisfy two requirements: "First, it must pass the truck-driver test: drop it out of a second-story window and if it survived, it might last two years. Second, it must be uncomfortable enough so customers would get up and move on." With so many different expectations of a single chair, it is no wonder that the perfect one has yet to be made. And what holds for the dining chair also holds for any number of other specialized chairs: folding, beach, lawn, easy, reclining, swivel, rocking, arm, club, desk, office.

The mid-nineteenth-century office, at least according to Charles Dickens, was furnished with desks at which one had to either stand or sit on a high stool, as did the likes of Bob Cratchit and Nicholas Nickleby, who spent long hours hunched over account books. Chairs were familiar pieces of furniture to the leisure class, but not to the lowly clerk, who stood to work at his desk all day. Today, this type of desk is rarely found in the office—save for that of the odd architect or engineer, the eccentric executive, or the writer with a bad back—even though some students of posture and anatomy assert that it is better to work at least part of the time standing up or perched on a stool than sitting in a conventional chair all day long.

By the end of the nineteenth century, the desk at which one sat had become a standard piece of office furniture for proprietors and managers, at least, and, naturally, it was accompanied by a chair. The rolltop's familiar orientation against the wall, its ogival support sides, which acted like blinders, and its broad Victorian expanse of cubbyholes forced the occupant to turn to his side or even completely around in his chair to talk with a visitor, client, or clerk. Such motion was awkward, not to say difficult, in a conventional four-footed chair, and so the swivel chair on wheels was frequently paired with the rolltop and other office desks. When alone, the office worker in a swivel chair had the freedom to turn effortlessly to reach even the desk's most remote drawers and compartments, not to mention to roll to files across the room. The chair's wheels accustomed workers to ease of movement, literally to a neighboring desk and metaphorically across the office floor and out the door to a new job.

The ability to lean back in a chair, whether to look up at a colleague or to put one's feet on the desk and daydream of another position, one with a larger and less cluttered workspace, made the tilt seat a very desirable feature. Well into the twentieth century, the most coveted work chair was one with wheels and with a seat that adjusted to the correct height and tilted and swiveled to all points on the office compass rose. To sit in such a chair, hands on its arms as if they were on the flight controls of an Airbus, was to be able to maneuver it like an airplane, yawing and pitching and sometimes even rolling over a sea of paper. The virtually indestructible oak desk chair is now a classic.

Unfortunately, wood burns, and the paper that accumulates in offices feeds the flames. Such concerns provided a ready market for metal office chests. So the Metal Office Furniture Company was founded in 1912 in Grand Rapids, Michigan, to manufacture fireproof safes and filing cabinets, which it called "steel cases." The company soon expanded its product line to include the Victor fireproof steel wastebasket and other accessories. A need to outfit the Boston Customs House with fire-resistant furniture led Metal Office Furniture into the fireproof metal desk business. Later, in 1936, it was manufacturing the steel desk and office chair that Frank Lloyd Wright designed for the S. C. Johnson & Sons Administration Building in Racine, Wisconsin. Metal Office made the steel table at which Gen. Douglas MacArthur and Japanese officials sat to sign the surrender documents ending World War II. After the war, the company introduced standard desk sizes, and in 1953, it began offering color options in office furniture. The next year, it changed its corporate name to Steelcase.

The Steelcase office chair, introduced at about the same time, did everything that the wooden office chair did, and offered more. The seat naturally pitched and yawed (and, like the classic oak, might also roll a bit when its mechanism aged), and it was adjustable in height and tilt stiffness. However, bettering its wooden counterpart, which was often customized with a soft but tattered cushion nestled into the scooped-out seat, the Steelcase chair was neatly upholstered.

The adjustable-height Steelcase office chair, dating from the 1950s, tilted, swiveled, and rolled.

For all of its classic features, the Steelcase chair also had its problems. But fortunately, the human body is amazingly tolerant of technological limitations and adapts to them. To adjust the seat height, the chair pilot had to get out of the cockpit and look under the plane. To adjust the back tension required a similar unchairing to get at the knob behind. Such little annoyances were hardly even noticed by most office workers, who mounted and dismounted their seats to test their adjustments and tweak them to their liking, but competing chair manufacturers were certainly aware of the shortcomings. The quest for the best chair has driven office-furniture designers as much as the search for a better mousetrap has driven basement inventors.

Today, a visit to an office-furnishings store provides ample evidence that designers and inventors alike have been busy indeed. Choices in office chairs have proliferated. An office worker now barely has to leave his seat to adjust the chair's height or change the growing number of parameters that fit the chair to his particular body. Pneumatic cylinders cushion the hardest landing. They make the climb fast and the descent fun, and everyone from boss to clerk has pretty much the same height-adjustment mechanism with which to play. But not everyone has the same chair.

The ultimate in office chairs at the change of the millennium was commonly believed to be the Aeron, manufactured by the office furniture–making firm of Herman Miller, which was founded in Zeeland, Michigan, in 1923. The fashionable Aeron looks like bare-bones technology, but it embodies the latest anthropometric, ecologic, and ergonomic features.

As is true with any redesign, the advertised advantages of the Aeron are best appreciated by recalling the disadvantages of the classic that it was designed to replace. The Steelcase chair had plastic-topped steel arms that were narrow and angular and thus hard on the forearms and elbows. The chair's arms had squarish front supports, which could press against the thighs. Its rigid arms could also dig into the occupant's

The Herman Miller Aeron office chair was introduced in the mid-1990s.

side when he or she leaned over or reached to retrieve papers from the bottom drawer. Armrest height could not be adjusted for individual users. In contrast, the armrests of the Aeron are wide enough to give broad support, and their rear-only mounting does not constrain the thighs. Armrest height is adjustable to individual needs, and armrests swivel inward and outward to follow the reach of the occupant switching between keyboard and computer mouse, and even points beyond.

The standard Steelcase's rigidly connected seat and back were upholstered, allowing little air circulation: Office workers' clothes stuck to the seat in steamy weather. Though the chair back was open at the bottom, the seat could still be as hot and sticky as that in a contemporary automobile in an asphalt parking lot under a hot summer sun. Furthermore, one chair size was supposed to fit all seats. But the Aeron's seat and back are contoured, like those in a sports car, and the fiberglass-reinforced polyester seat and back frames are of skeletal proportions, across which a mesh fabric is stretched. The fabric, which conforms to the occupant's body and through which it can breathe, allows air to circulate and thus provides some relief during the hottest and most humid summer months.

Since the Steelcase's seat height and tilt stiffness controls required a chair jockey to guess how much to turn them before getting back in the seat, few office workers fully mastered these controls and so resigned themselves to suboptimally adjusted chairs. With the Aeron, the seat and back adjust individually. Seat height, tilt, and spring tension can be customized with distinct and easily reached knobs and levers. Arms and lumbar support are also adjustable. Adjustments are made by handy controls, and illustrated instructions for operating them are only a mouse click away on Herman Miller's Web site. In the new office without walls, where every day is a casual-dress day, the egalitarian Internet leads executive and secretary alike through the same motions.

On the Steelcase chair, narrow casters were susceptible to locking in the wrong position for rolling. A cruciform base allowed casters to be positioned well within the edges of chair, which provided insufficient support and allowed an occupant to topple over if leaning too far forward, sideways, or back. The large and free-turning casters on the

Aeron allow this chair to be rolled effortlessly in any direction. A now-familiar five-footed base provides considerable stability against tipping over, even when the occupant leans way back in the chair.

The Steelcase chair was, of course, made of steel. Other materials included the plastic, vinyl, and foam rubber that were once de rigueur in office furniture. Fashion and function, but not recyclability, governed choices. The ultimate fate of the chair and its materials, and their impact on the environment, was not an issue in the 1950s. The Aeron is advertised to be environmentally friendly. The chair is made mostly of recycled materials, like aluminum and polymers, and its parts themselves are said to be 80 percent recyclable. According to Herman Miller, less energy is consumed in manufacturing the Aeron than is used for chairs with foam and fabric upholstery. The Aeron chair is even shipped in returnable and reusable packaging.

The success of the Aeron, launched in 1994, did not go unnoticed by competitors. Steelcase, which discontinued its near-indestructible chair in 1973, came out in 1999 with the Leap chair, which has many of the same adjustment options as the Aeron, but not its patented features nor its cachet. Instead of mesh fabric to keep the occupant cool, for example, the Leap has slits in its seat bottom and back.

The modern office worker's heightened expectations for the latest-model task chair must certainly be due at least in part to experience with the newest automobiles. Few office workers at any level today would undertake an all-day trip in a fifty-year-old sedan with a bench seat, whose only adjustment is its distance from the steering wheel. (Some students of the chair, like the architect Galen Cranz, have nevertheless asserted that the bench seats found in 1950s cars are still the best automobile seats around.) Most of us have become accustomed to bucket seats with power adjustments within easy reach, with some seats even remembering several drivers' favorite configurations. Going to work in the morning may not be quite like setting out on an eight-hour car ride, but it can be just as daunting. The worker who emerges from an optimally adjusted bucket seat in a late-model car is not likely to want to spend the workday sitting before a computer screen in a half-century-old chair.

Historically, what made any chair difficult to design was the fact that one size had to fit all, even though not all users could fit comfortably into one size. Adjustment features have helped, but the typical office chair still has only a limited range of seat and back height, seat depth, tilt, and other adjustments. Even the Aeron, which like the chairs in the house of the Three Bears comes in three basic sizes, can provide only a limited range of adjustments with which a working girl can seek the configuration that is "just right." But a real person in a real place is not a Goldilocks in a fairy tale, finding the perfect chair in a strange place. In fact, for many people finding the perfectly comfortable chair anywhere is a never-ending quest, which is not surprising, since finding perfection in anything designed is a virtually impossible goal.

The basic problem is, of course, that, like bears, people come in different sizes and shapes. The range of people's sizes and shapes was given currency by the industrial designer Henry Dreyfuss, who was a pioneer in using ergonomics for seating, among other things. His firm, relying heavily on data gathered by the military, developed the first edition of the indispensable resource book in the field. Initially published in 1960 as *The Measure of Man,* but now titled *The Measure of Man and Woman,* this handbook contains anthropometric charts showing the normal range of dimensions of the parts of the human body that must be accommodated by everything from chairs to potato peelers to electronic devices. With the aid of resources like the Dreyfuss book, it is possible to design chairs that should fit the physical dimensions of the fiftieth percentile man or woman (but not possibly both) "just right," and should accommodate the dimensions of about 98 percent of able-bodied men and women. (The latest edition of the handbook also deals with ergonomics for people with disabilities.)

The normal distribution of heights of adult males, according to one of the graphs in *The Measure of Man and Woman,* is centered about 69.1 inches, with the first percentile cutoff at 62.6 inches and the ninety-ninth at 75.6 inches. This gives a height range of over a foot that a one-size-fits-all chair must accommodate. In addition, there are the wide ranges of bottom widths, hip-to-knee distances, and, important for

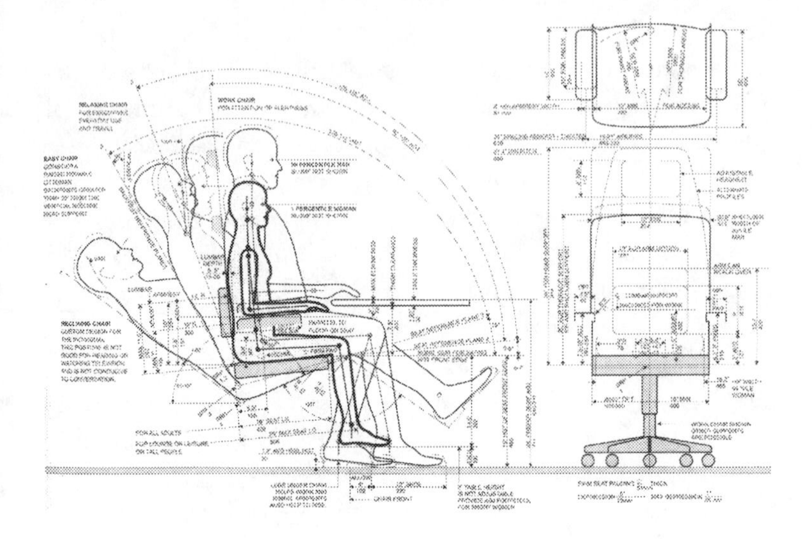

The range of people's sizes and shapes forces compromise in the design of everything from office chairs to table tops.

armchairs, shoulder-to-elbow distances. The permutations of the combinations that can come into play in the course of designing a chair are staggering. It is no wonder that while some shorter men might find their feet dangling from a chair, some taller women sitting in the same chair will find their knees at a higher level than the seat. The situation is not so bad in adjustable automobile seats, of course, but for those of us pushing the adjusting mechanism to its limits, a long car ride can become an uncomfortable one.

Naturally, those at and beyond the limits of design learn to accept and adapt to the generally average (idealized) size of the world of made things. The short become accustomed to sitting forward in their chairs, so that the front edge of the seat does not cut into their calves. The tall sit back to get as much support as possible beneath their thighs. As uncomfortable as we might be, we do ride and tame the average chair. And though fitting the physical body is not the same as fitting the fashionable or psychological expectations of the mind, the fit is usually close enough to excuse the furniture designers, for they can do only so much.

My son is a couple of inches taller than I, but he has more headroom in our car because he is longer in the legs, while I am longer in the torso. Yet, as much as two men differing even by more than a foot in height can drape themselves over the same standard chair without being overly conspicuous, they cannot at all hope to wear the same average-size clothes without looking a little foolish. The lucky few may be able to keep themselves in custom-tailored suits, shirts, and shoes, but most men must outfit themselves with off-the-rack and off-the-shelf clothing. Men's ready-made shirts do come in a range of neck and sleeve sizes, of course, with the former usually available in half-inch increments and the latter in full-inch ones, but this concession by clothing companies to the normal distribution does not mean that all men are happy. It is largely the luck of the draw whether one's arm stops growing exactly at (or even comfortably close to) integer length. I figure that my sleeve size is midway between thirty-five and thirty-six inches, and so I have the choice of wearing my sleeves either short or long. (Some manufacturers do make shirts with sleeves designated 35–36, which fit me better, but their versatility usually depends upon cuffs with two buttons, leaving me to choose between a too-tight or too-loose fit at the wrist.) If shirts were readily available in actual half-inch sleeve increments, arms like mine might more easily be fit "just right," but the price would likely increase for the customer, as would the size and complexity of a store's inventory.

Making and stocking chairs with seat sizes even in two- or three-inch increments would be a nightmare for furniture stores. And what if chairs were made in different seat heights? What size chairs should a restaurant have? Should a family buy different sizes of chairs for its dining room? How would they function around a table of fixed height? Should they all have the same total back height, so that they would appear to be a set? What would they look like when arranged against the wall during a reception?

Except for the fortuitous fit, if there ever is to be a near-perfect product of design for human use, it most certainly will have to be a custom-made thing for a single individual. (Computer-driven custom manufacturing of everything, which may be an option in the future, is not practical at present.) But can even a custom-made suit fit equally well in all attitudes and situations and climates in which the wearer is likely to find himself? It is possible to order a custom-made chair from some craftsmen, but to what guest could such a chair be offered? It has been proposed that once a worker does find a chair that fits "just right," that chair should be moved from job to job along with the owner's framed family photos. But how would interior designers accommodate an unpredictable future assortment of chair styles, colors, and fabrics as managers moved up the corporate ladder?

For all of its limitations, the generic chair is still a wonderful design achievement. But few masterpieces of chair design, which have become icons of high consumer culture, can withstand close scrutiny. The Barcelona chair may be a classic of aesthetic design, but it is impossible to get up from gracefully. The competing multifarious objectives of design are too diverse, too diffuse, too demanding, too contradictory. There is no way that any design can satisfy all ideals and all real people simultaneously; there is simply no perfect design. All designs must involve trade-offs, if not in materials, then in function; if not in cost, then in fashion; if not in quality, then in proportion; if not in size, then in shape; if not in this, then in that.

The object of industrial design is not solely to make unique objects as close to aesthetic perfection as humanly possible. That is art. Practical design has to deal with the constraints of science and engineering as well as of art, with money and manufacturing, with use and wear, with form and function. There is no avoiding the fact that these various constraints are often in conflict, sometimes in direct conflict, and so there must always be compromise in design. There must be compromise by the manufacturer in what is made and how it is made. There must also be compromise by the consumer in what is bought and how it is used. There must be compromise on compromise. It is in the nature of design. All engineers and designers know they always must compromise, whatever the situation.

So central is this idea to engineering that it has on occasion been defined as "the art of compromise." Whether designing a chair for a new millennium or a gadget for the quotidian, engineering and design teams have had to and will always have to wrestle with the inchoate mass of possibilities and contradictions and perceptions. A 1960s study of chair use concluded that a person given more than two adjustment options might not be able to remember tomorrow which settings most closely reproduce the ones that made the chair "just right" yesterday. As our ideas of fit and comfort evolve with time and situation, so do our ideas of design success and failure.

THIRTEEN

Familiar Design

THE MOST COMMON things in the world are often invisible to us, and thus also is their design. It is not that we do not see these things, for we must see them at least peripherally in order to use them. In fact, we see them so routinely that in themselves they hold no interest for us. They are familiar, ordinary, and all too common. They are simply the means to an end. We use them, but we do not look at them. They are so predictable in their form and function that we do not give them a second thought. We grab them, use them, and let them go. They remain as we found them, ready for another encounter, perhaps of some great moment but of no great remembrance.

In my home, I count sixteen doors with knobs, which occur in pairs. (Though I don't know why there needs to be a knob on the inside of the linen closet's door, since I can't imagine even a child being able to squeeze into it and closing the door behind.) These are all conventional doorknobs, and they are cleverly designed, in that they work equally well by being turned to the right or to the left, thus making them devices truly suited for both right- and left-handers. Knobs also appear by design to be installed at a standard height on each door. In my house, that height is three feet, give or take a fraction of an inch. I have become so accustomed to this height that I can approach a door, grasp its knob, turn it, and open the door in one fluid motion. My arm automatically reaches out midstride to dock with the oblate spheroid, turn it just enough to release the latch, and push or pull with a force that moves the door into a position that gives me the necessary opening to pass through.

We become so accustomed to actions like opening doors in our own home and neighborhood that we can become confused in another home, town, or culture. On a visit to England's Cambridge University, I was put up in Peterhouse, the college right next door to the Engineering Department. I was introduced by my host to the beadle (college custodian), who showed me to the entrance to the building in which I was to stay. The location of my rooms described and my breakfast order taken, I was left on the walkway. As the beadle went back to his post, I climbed the few steps to the front door and grabbed its knob. I turned it and pulled, but the large door would not budge. I had been given no key, but I assumed the door was locked, and so I retraced my steps to the beadle's counter and reported that the door to the building was locked. He informed me that it surely was not, and he trudged back there with me. He led me up the stairs, grasped the knob, twisted it, and pushed the door wide open. I would like to think it was jet lag that had me pulling on a door that should have been pushed. Being so used to houses in America with screen and storm doors that had to be opened out before the front door proper could be reached and opened in, I had instinctively pulled on a door that should have been pushed. Doorknobs give no hint as to which way the door opens, but I should not blame an inanimate thing for my presumption to pull in a single-door culture where I should have pushed.

National differences are no excuse, either, for jumping to conclusions about the anomalous positioning of doorknobs. Our family once stayed in an American bed-and-breakfast that had abnormally low doorknobs, so abnormally low that they called attention to themselves. I had to stoop down to open doors. The cut-glass knobs were attractive antique models, as were the finishings and furnishings of the entire house. But almost everything else about it suggested height: high ceilings, tall doorway openings surmounted by high transoms, tall windows. The present proprietor could not explain why the doorknobs were so low, other than to remark that he thought people were shorter two centuries ago, when the house was built. But I do not recall being in any other nineteenth-century American building with such low doorknobs. To me, this house was an aberration, apparently an idiosyncratic design built the way it was for reasons no longer known.

I expect that how high doorknobs were to be located is recorded in some old carpenters' manual, but I have not yet found it. Nor have I found such information written down in the instructions for installing modern doors and their knobs. One extremely explicit set of step-by-step instructions begins by saying, "Open the box that your new lock is in," then instructs the do-it-yourselfer how to use a paper template to establish the location where a hole must be drilled through the door. The horizontal distance of this hole from the edge of the door is critical if the knob's spindle assembly is to engage the latch correctly. The instructions give the precise distance of the hole from the edge of the door ("2-3/8″ exactly"), but there is no mention whatsoever of how high up from the bottom the hole should be drilled. Another set of instructions, this for installing folding closet doors, likewise neglects to say how high the doorknobs should be. We are, presumably, expected to know this traditional dimension, which evidently has nothing to do with whether the house is occupied by short or tall people. (The design of panel doors does, of course, provide some guidance for locating the doorknob, but the unbroken flat design of plain modern doors gives no hint.)

I am definitely taller than average by today's norm, and yet I feel that the standard three-foot-high location of doorknobs is just right for me. I have not heard any of my shorter acquaintances complain that doorknobs are too high for them. Nor have I heard any seven-foot basketball players complain that ordinary doorknobs are set too low for them. (They seem to learn early on to adjust to a world of short and low objects.) In fact, everyone appears to adjust easily to the standard height of doorknobs and to think they are just about where they should be. This should not be surprising, for the height of doorknobs stays constant, while we grow from little tykes reaching up for doorknobs to gangling teenagers reaching down for the same knobs. Our height changes are slow enough for us to adjust and adapt, and so we think at any time in our development that the doorknobs are located in about the right place. Whatever the reasons for doorknobs being located three feet off the floor, people whose heights range from two to seven feet can feel these knobs are just about where they should be because they are just where people remember them always to have been. And because they have been where they have been for so long, they have become virtually unnoticed in our daily use of them.

Perhaps another device as equally invisible and as frequently used as the doorknob is the light switch. In fact, beside almost every door with a doorknob is a light switch, and in some cases, we use it every time we use the door. It may at first appear curious that light switches are not placed at the same height as doorknobs. If we are accustomed to reaching for doorknobs in a certain location, would it not have been sensible to locate the light switches at the same height? Since our hand is already used to operating at doorknob height, would it not be most convenient just to swing the arm to the side of the doorjamb to flip the light switch on or off?

Without doubt, there would be a certain symmetry and order to having the doorknob and light switch at the same level, but the opportunity for that to happen is long past. Doors and doorknobs clearly predated the introduction of electricity into houses. In fact, older houses had to be retrofitted with electricity and its accoutrements. The oldest light switches I have encountered consist of a pair of push buttons arranged one above the other. Depressing one button (usually the top) turns the light on and at the same time pushes the other button out, ready to be pushed in to turn the light off. Button light switches can still be found in some old houses that have yet to be renovated, and if my memory serves me correctly, they are more or less at the height that we now consider standard. Why that height was selected in the first place is a question whose answer can be speculated upon.

When electric lighting was a new phenomenon, it was also, naturally, the object of much attention. Among the earliest electric switches were knife switches, the kind mad scientists have been known to throw in order to drive life into monsters. Such bare technology was not refined or safe enough for the home, and so the more genteel push-button switch, its mechanism concealed behind a decorative faceplate,

We have become accustomed to doorknobs and light switches that are located at predictably different heights.

was adopted. People then, no doubt, pushed the on-off buttons with deliberateness and probably not a little wonder.

When it came time to install electrical switches in a house retrofitted with electricity, the location of the preexisting gas fixtures, conduits, and controls were usually exploited. Gas valves were typically located high on the wall, if not on the fixtures themselves, and so electric switches followed suit. Even if the location of electric switches had been thought through from scratch, the action of pushing buttons would likely have dominated any decision about location. It was also probably considered appropriate to locate the pair of buttons high enough on the wall so that they, like gas controls, were out of reach of small children. After all, there was a certain ambivalence and concern about the safety of electric lights, and especially when the current that flowed through switches into them was of the alternating kind. (Thomas Edison, who favored direct current, recommended the use of alternating-current generators made by his rival George Westinghouse when electrocution was being considered in place of hanging for carrying out capital punishment.) Furthermore, the action of pushing the right button would be performed more easily by adults if the buttons were not too high for the hand to reach or too low for the eye to see. Reaching down to doorknob level to push a button into the wall would not have resulted in a very natural finger movement. When the flip switches familiar to us began to replace the push-button variety, it made sense to install them at the same height as the buttons. After all, that was where the hole in the wall already was and that was also where the wires terminated. And that was where people had come to expect light switches to be.

Today, it would take a mischievous electrician, an idiosyncratic home builder, or an extreme ergonomist to install light switches at the same level as the standard doorknob. As logical and symmetrical as it might seem, the radical design change would look wrong, feel wrong, and be wrong. The location of light switches at about four feet off the floor has become as "right" as the three-foot height of doorknobs. Also, the location of the switches on the wall is as natural looking as the location of one's ears at about eye or nose level, not at mouth level, on one's head. Custom, tradition, and expectation are as much a part of design as aesthetics. In fact, our aesthetic sense is as much determined by what we are accustomed to as by what would appear to be logically right.

It is also the case that most light switches are operated by a person entering a room as well as by one exiting it. While a lower light switch would be convenient to throw by someone on the same side of the wall as the switch, it would not be so convenient for someone entering the room from the outside. The arm reaching around the doorjamb has a better range for the higher position than the lower one, no matter how sensible the latter might at first seem. In fact, light switches located at upper-arm height—a compromise between doorknob and eye level—are reached equally easily from both inside and outside the room. There is, not surprisingly, a sense to it after all.

When a new house is built today, as soon as it is framed in wood (or, increasingly, in metal), the electrician is brought in to wire it. He will install electrical boxes for light switches beside door openings that have no doors and therefore no doorknobs. If it is not obvious, he will have been told or otherwise informed on which side a door will open and will install the box for the switch on that side. He will not need to know how high the doorknob will be, for that is essentially a given and is not relevant to the control of electricity anyway.

Unless directed to do otherwise, the electrician will install the boxes for outlets a foot off the floor, at reasonable locations around the room, using, in the absence of any other guidance, a rule of thumb to determine the number to include. He will use his judgment to make choices and will compromise within the constraints of any problem or regulation, including the building codes that must be observed. Most home builders leave such design decisions to the electrician, because they are not things that are given a great deal of thought by most people. In fact, the typical electrician does not give them a great deal of thought, either. He just wires the house the way he has learned to wire houses, unless he is dealing with one so unusual that it does not remind him of any previous situation or unless someone dictates the details and thus defies the standard design.

The electrical outlets in our home are by and large located at the standard height above the floor. However, they are wired in what at first seemed an unusual way but which eventually has proven to be a most convenient one. The outlets look ordinary, comprising a pair of receptacles stacked totem pole–style one atop the other. In rooms that were rewired when the house was renovated, all upper receptacles are controlled by the room's light switch, so that lamps plugged into them can all be turned on with a single flick of the switch. All lower receptacles are hot-wired, however, so that anything plugged into them will not be turned off when the light switch is. This is especially convenient in my study, where I can turn the lights on and off without affecting my computer.

As I have with doorknobs, I have gotten used to the conventional height and operation of the familiar electrical outlets and light switches in my life. For the light switches, I know instinctively just about how high to reach inside a darkened room. That is not to say that I know how far beside the door to reach. For whatever reason, usually having to do with how the door and wall are framed with two-by-fours, the horizontal placement of light switches relative to a doorway does not have nearly the same consistency as the vertical. Most of the switches in our house are within a few inches of the molding around the doorway. However, there are a few that range farther than that, with some being as much as a foot from the door. These are the ones I often have to fumble for, though fumbling soon reminds me of the anomalous placement and I quickly zero in on the switch. In spite of the irregularities of a house of unique additions and revisions and handyman repairs, before long I learned to find my way around it in the dark.

Even where there is a pair of switches behind a single wall plate, we quickly learn which switch to throw to turn the desired lights on or off. This is the case in my study, where one switch controls the wall outlets powering the desk and floor lamps and the other the recessed lighting above the bookcases. But in other rooms I have to stop and think. In our kitchen, there are three switches behind a single wall plate, one controlling the recessed ceiling lights over the counters, one the lights under the cabinets, and the third the light in the pass-through to the hall. The switch for the pass-through is between the two switches for the kitchen lights, an illogical arrangement, although the kind of inconsistency that would not surprise Donald Norman, who is an eloquent spokesman for the logical design and placement of light switches, door handles, and other everyday things. Yet, after living in this house for ten years, I still sometimes flip the wrong switch when I want to turn on one set of these lights. Even when I am looking directly at the triad of switches, sometimes I have to think which is which. A panel with four switches inside the front door gives me even more trouble, controlling as it does the downstairs and upstairs hall lights, the porch light, and the driveway lights. I attribute such personal lapses to the triviality of the task and the lack of attention I generally give to flipping a light switch.

Several rooms in the house have two-way switches, which present another kind of problem. These control the lights in a room that can be entered and exited from different doors. The light over the stairway is also connected to a two-way switch, so that we can turn the lights on downstairs and off upstairs, and vice versa. The arrangement is as convenient as the design concept is brilliant. However, the scheme also means that the position of the switches is not always consistent with the condition of the light. Depending on how the switches were installed in the first place and how they were last operated, one or the other of them can be in the down (off) position when the light is on, or in the up (on) position when the light is off. This can be disorienting, but who would not gladly accept a moment's confusion if it means being able to turn the staircase light on without having to go downstairs in the dark? The convenience and safety seem well worth the compromise in consistency.

In two of the rooms in our house, the light switch is paired with an electrical outlet behind the same plate. One of the rooms is a bathroom, in which case the outlet is conveniently located near the vanity. The other switch/outlet pair is located in a large storage closet, where the outlet provides a convenient place to plug in an extension cord in a room whose conventional wall outlets I have rendered inaccessible by placing boxes and other stored items in front of them. However, in both of these rooms, in order to accommodate the outlet below it, the light switch operates horizontally rather than in the customary vertical way. It thus sometimes takes a special adjustment in reflexes and a contortion of my hand and mind to turn on the light without trying to flip the switch up. (That these electrical boxes were not installed horizontally is a testament to the habit of design and the dilemma of installation. If the box had been installed horizontally with standard equipment, the switch might be in a conventional orientation, but the outlet would not.) But my annoyances with the arrangements are minor and perhaps attributable to the fact that I enter these rooms much less frequently than most others in the house and so have not yet developed a practiced feel for them.

As is usual in most homes, in ours some of the doors into rooms open from the left and others from the right, and so the light switches can be either to the right or the left of a doorway. However, my tendency is to use my right hand to turn a doorknob, regardless of which side the knob is on. My hand has grown accustomed to the small space between the knob and the jamb, and I have learned to keep my fingers close to the knob so as not to wedge them into that space. But during a stay in a hotel in Brisbane, I did get my fingers squeezed—and more than once—between the knob and the jamb of a closet door. The handsome knob, which was in the shape of an ellipsoid, elongated in the vertical direction, had been installed so close to the edge of the door that a surprisingly narrow space was left between the knob and the door frame when the knob was turned. Furthermore, the knob required an inordinately large amount of twist to release the latch, and so my fingers were naturally wedged into the ever-narrowing space created by the rotating geometry. None of the other doors I encountered in Australia had this feature, and so I attributed it to an attempt to design such a strikingly different doorknob that all conventions were ignored—or forgotten. Some of the most egregious mistakes in design are a result of an excessive attempt to distinguish something common from the commonplace.

We become so accustomed to where pieces of hardware like doorknobs and light switches are located, and how they feel and function, that even the slightest departure from the expected and standard can cause confusion, inconvenience, or discomfort. It might make a lot of sense to locate all wall switches on the outside of the room whose lights they control. In this way, the switch would be in plain view and the lights could be turned on before entering the room (though contortions might be required on exiting it). In fact, it would be even more convenient to locate two light switches side by side next to each door connecting two rooms. In this way, the lights to the room being entered could be turned on and those in the room being exited could be turned off at the same time, thus also conserving power. Many more switches would have to be installed in each house, of course, but the potential overall energy savings could be enormous. Motion sensors, which have come to be used increasingly in modern office buildings and homes, accomplish much the same thing.

Locating light switches (or motion sensors) outside the doors of hotel rooms would not work, of course, because then the room lights could be capriciously controlled by any prankster walking down the hall. As things are, entering a strange hotel room and trying to find the correct light switch, or even any light switch at all, can be a challenge. There seems to be no standard location for the switches in a hotel room. One cannot even count on finding a switch just inside the door, on the wall that is in the same plane as the door, because space is at a premium in hotels and there is often not enough room in the entrance corridor for that wall space to exist. This usually means that the light switch is on the wall perpendicular, but how far into the room can vary from a few inches to a foot or two. Even when the switch is found, it is often ineffectual, controlling either an outlet into which is plugged a lamp whose bulb is switched off or a lamp that has been fitted with an energy-saving bulb that takes a second to light up. Being used to the incandescent bulb's instantaneous response to a light switch, a hotel guest may immediately throw the switch back off and look for another one to control the lights. Thoughtfully designed hotel rooms—preferably equipped with three-way incandescent bulbs—often have light switches with illuminated toggles or master switches that control every light in the room.

As much as we can adapt to new configurations of technology, we retain our psychological expectations of how things work—or are supposed to work. We accept a considerable number of essentially arbitrary traditions, conventions, and seemingly capricious compromises in exchange for the convenience of not having to think about such things as how high a doorknob will be or which way a light switch must be flipped. As long as they are located pretty much in the vicinity of where we have become accustomed to finding them and as long as they operate in pretty much the way we expect them to, we can use them with our eyes closed. Even when designs are not perfect, they can be both effective and admired.

FOURTEEN

Design by the Numbers

WHENEVER SOMETHING is designed, decisions have to be made. The decisions may not necessarily be predetermined, though they may follow from a fundamental design philosophy, which itself represents another order of choice. Design decisions are necessary because deliberate progress cannot proceed without choices—as to whether one part goes on the right or the left of another part, whether a component is larger or smaller, or whether it is round or rectangular. Such decisions, arbitrary as they may at times seem, often have profound effects on the way technology develops.

In early conceptual designs for space stations, such as those drawn in the 1940s and 1950s by the imaginative artist Chesley Bonestell, the structures were often large and wheel-like in shape. The steady rotation of the massive wheel and the accompanying centrifugal force were meant to simulate a gravitational field, and hence the sensation of weight in an otherwise weightless environment. At the time, this seemed like a desirable feature for a space station, so that astronauts could work under conditions like those on earth as much as possible. When it came time to design and assemble real space stations, however, they were made boxy and long and did not rotate, thus forcing their occupants to deal with weightlessness. Why?

Building any kind of major structure in space requires great amounts of construction materials to be carried into orbit. The size of the space shuttle's cargo bay clearly limits the size of the parts, and so building a giant wheel would have required a great number of expensive launches. A partial wheel would not easily have provided an environment with artificial gravity, and so the astronauts would likely have had to work in weightlessness anyway until the wheel was completed and spinning. By abandoning the goal of artificial gravity, rectilinear structures like the International Space Station could be constructed in a modular fashion and be operational, if on a smaller scale, well before they were fully completed. Besides, one of the main advantages to having a space station in orbit is to conduct experiments under conditions of microgravity, something that would not have been possible in a rotating wheel-like environment. In any situation, what at first might seem like an obvious and desirable design feature often undergoes reconsideration, and the design thus undergoes considerable change as new constraints and objectives become more clearly defined and understood.

In the case of highly visible artifacts, like the space station, there is often a clear historical record that documents for us the evolutionary design and decision processes that brought us to the thing that we have, or the thing to us. Looking into such a record in detail can help us understand why certain things are the way they are and, perhaps more important, how things in general get to take the form that they do. One case study that has these attributes encompasses the keypads of two familiar artifacts that most of us use daily: those of the telephone and the handheld calculator.

In an earlier book, I posed the question of why these two common devices have different layouts for their push buttons, the telephone's keypad having the numbers 1,2,3 on its top row and the calculator's having 7,8,9 in the same place. The purpose of the question was to encourage readers, especially students, to look closely at two items that they encounter every day and to think about and speculate on what reasons there might be for the two different solutions to essentially the same design problem—inputting numbers—to have developed and persisted. Readers were not expected to delve into the history of keypads, but, rather, to reflect upon a curious example of a lack of standardization in the world of invention and design and whether that lack

The arrangements of the numbers on push-button telephones and electronic calculators are curiously different.

has serious consequences for users. Perhaps the time has come, however, to reach into the historical record.

The most commonly given answer as to why the two keypads are different is that the modern calculator and the modern telephone derive from different technological roots. The calculator's ancestors are cash registers and calculating machines, on which the keys were arranged vertically in ascending order, from 0 at the bottom to 9 at the top, an order that expressed visually their ascending magnitude. Push-button telephones, on the other hand, derive from rotary-dial phones, which, in addition to digits, had to incorporate letters of the alphabet—naturally arranged in alphabetical order, reading, as one would an English text, from left to right and from top to bottom. There is a nugget of truth in this explanation, but the full story is a bit more complicated.

The technological foundations for a rotary-dial telephone date from 1891, when a patent was issued to Almon Strowger, a Kansas City, Missouri, undertaker, who is said to have become suspicious that telephone operators were accepting bribes to direct calls intended for his mortuary to those of his competitors. He invented an "automatic telephone exchange" that eliminated the need for an operator, at least for local calls. Initially employing buttons (and thus anticipating the push-button phone), by 1905 Strowger's scheme had evolved into one using a dial wheel. The American Telephone & Telegraph Company licensed Strowger's invention in 1916, but it was years before rotary-dial telephones were widely used throughout the Bell System.

The standard rotary dial was a wheel with ten finger holes arranged in an almost complete circle, with most of the holes labeled with a triplet of letters in addition to a number. Dialing a letter or numeral was done by inserting the index finger into the appropriate hole, turning the dial wheel clockwise to the finger stop, and releasing the dial to allow it to return to its rest position, ready for the next digit to be dialed. The use of the rotary dial may be obvious to those of us who once used such telephones daily, but an engineer who collects the devices has found their operation to be less than transparent to the uninitiated. According to the engineer, "one 10-year-old boy who was trying to dial a three put his finger in the zero dial hole, brought it to three, and then released it." The zero hole was, of course, also labeled OPERATOR, and the boy may have drawn the seemingly logical conclusion that it was the hole that he, as the operator of the phone, was to use to drive the device.

When the rotary-dial phone was still a familiar fixture in offices and homes, women with long or recently polished fingernails dialed by using the eraser end of a pencil or a special plastic dialing aid, which had a knob similar to the top of the Bic four-color pen. Whether driven by a finger or an extension of the finger, the spring-loaded wheel clicked when turned and then returned on its own with a characteristic ratcheting sound to its home position as each number was dialed, a loud and slow process compared to pushing the buttons on today's Touch-Tone telephones.

The finger holes were arranged around the dial wheel in ascending order counterclockwise. The number 1 and the letter unit ABC (paired with the number 2) took the least amount of time to dial, since they required the least distance to rotate and then return to their rest position. (This is why the largest cities, like New York and Los Angeles, which were expected to be most frequently called long-distance, were assigned area codes like 212 and 213.) The combination of letters and numbers on a rotary dial was necessary because older telephone numbers comprised named and numbered exchanges, which served as prefixes for four-digit numbers. The mystery movie *Butterfield 8* (and novel by John O'Hara) derived its title from the telephone exchange BUtterfield 8, which was connected to by dialing B-U-8. (Dialing four more digits reaches a line within the exchange.) In New York in the 1950s, a telephone call was made by dialing a complete seven-digit phone number, such as MU 5-1234. The MU indicated that the telephone being called was located in the Murray Hill section of the city, which is the area in which the New York Public Library now sits, and before that a reservoir holding water brought in by the Croton aqueduct. (The hyphenated combination of alphanumeric exchanges and four-digit numbers has come down to us in the way most of us group, pronounce, and remember the purely numerical numbers of today—for example, 685-1234, rather than 6851-234.) That today's push-button telephones have their keys arranged with 1,2,3 (along with ABC and DEF, the 1 key never having been paired with letters of the alphabet) comprising the top row has more to do with human-factors research than tradition, however, as the historical record demonstrates.

Human factors is the shorthand name for human-factors engineering, which is also known as human engineering and engineering psychology. This field of specialization deals with the human-machine interface and considers the human user to be part of a person-machine system. Among the principal goals of human-factors engineering has traditionally been to contribute to the design of instrument displays and controls and to the efficiency, safety, and reliability of the use of machines. The field is a part of the larger one of industrial engineering and traces its roots back to the scientific-management movement spearheaded by Frederick Winslow Taylor in the early twentieth century. Research in the field was greatly intensified during World War II, especially in Britain, where it became known as ergonomics.

The objects of human-factors studies can range from the size, shape, and arrangement of the dials, buttons, and levers in an airplane's cockpit or a nuclear power plant's control room to the effect of the diameter of pencil lead on the response time for filling out the answer sheets on multiple-choice exams. When in the 1950s AT&T's Bell Laboratories began looking seriously at the use of push buttons on telephone sets, the buttons' spatial arrangement, physical action, and ease and reliability of use became natural subjects for study by human-factors engineers and psychologists.

Push-button devices were nothing new. As we know, early light switches were operated by push buttons, some of which are still in use in older homes. Automobile radios with push buttons were also common by the 1940s, and I recall our family's first car, a 1948 Dodge, having them. Such push-button devices were, however, principally on-off switches, whose function was indicated by position, and they were not used for extensive data entry, such as of a series of numbers designating a telephone line. Key sets, on the other hand, consisted of a matrix arrangement of keys or buttons, each corresponding to a unique digit. These also were not new in principle, for cash registers and mechanical calculators had long used multiple columns of numbered keys, one column for each decimal place. By the early 1950s, key sets of the kind we might recognize today were in use on coding devices, computers, and communications equipment. However, as a 1955 article in the *Journal of Applied Psychology* noted, there appeared to have been "few systematic studies concerned with the design factors that make keysets easy or hard to use."

The study that prompted this remark was motivated by the fact that long-distance telephone operators were making errors in entering numbers like 815 RE 4-0267 when using a ten-button key set arranged as follows (with the letters of the alphabet located on the usual keys):

4 5
3 6
2 7
1 8
0 9

According to the researchers, the patterns of errors associated with using this key-set arrangement suggested that a different ordering of the keys might help reduce the error rate. As a first step in seeking the best arrangement of the alphanumeric keys, the researchers decided to "find out where people say they would *expect* to find letters and numbers on six different key-set configurations"—namely, ones arranged as follows:

The first four arrangements were also the top four choices, by frequency, of the subjects of the study, and they arranged the numbers (and corresponding letters) as follows:

1 2
3 4
5 6
7 8
9 0

1 2 3 4 5
6 7 8 9 0

1 2 3
4 5 6
7 8 9
0

0
1 2 3
4 5 6
7 8 9

The leftmost of these arrangements, incidentally, would be the one that Martin Cooper, an electrical engineer with Motorola, would use in

1973 to make the first cellular telephone call. The handheld device that he used weighed some two and a half pounds and was about the size of a box of tissues. Over a decade before that, however, the "most obvious finding" and perhaps not surprising result of the study was "that people arrange numbers and letters in order in which they normally read." The researchers commented on the fact that, of several contemporary "calculating devices" that they were able to look at, only the IBM punch-card machine used an arrangement of keys that the study found highly preferred—namely, the one on the far right in the four arrangements shown here. The arrangement second from the right was found on the multiplier keys of the Friden calculator and on the key set of the Remington Rand adding machine. As for "most other calculators," they had "keys reading upward in vertical rows of ten," with each row designating a different decimal place, as on cash registers, with the 0 located below the 1, a placement seldom found preferable in the tests. (The electronic calculator had not yet arrived, of course.) The researchers admitted to having "no assurance that these differences between calculator keysets are of any practical importance," and considered their study "only a first step toward finding the keyset which would give the fewest errors and shortest keying times." Thus was the quest begun.

Those conducting human-factors engineering studies on the design and use of push-button key sets admitted that "the number of possible key arrangements, force-displacement characteristics and button tops is very large—too large to be tested. A selection of characteristics was, therefore, made on the basis of prior knowledge, user expectation and broad engineering requirements, so that we could examine only the region around an expected maximum of performance."

Such a situation is not uncommon in engineering design, whether it be a problem of choosing the material for a nuclear reactor's core, the type and style of bridge for a particular crossing, or the arrangement of the keys on a telephone set. Engineering judgment, insight, and simple common sense contribute to narrowing down the seemingly limitless field of choices to a manageable few that can be compared in a rational way in a reasonable amount of time to provide the basis for reaching a final decision.

By the late 1950s, field trials of push-button telephones were under way, but research continued at Bell Labs to determine the optimal arrangement of the key set, as well as the size, spacing, and tactile characteristics of the action of the buttons themselves. Among the arrangements of buttons that had been tried, in addition to the variety of rectangular arrays, were circles, triangles, and a cross pattern. It had also been established that a seven-digit telephone number could be keyed in as much as five times faster than it could be dialed in the then-conventional way. Until I had tried my first push-button telephone, I thought dialing time to be unimportant. However, after getting accustomed to operating a new push-button telephone installed in my office in the mid-1970s, our old rotary phone at home seemed frustratingly slow to dial. Today, some cheap Touch-Tone-style phones cannot keep up with the speed that our fingers have learned to achieve in using a keypad. Such a failure of technology to keep pace with the people who use it is a driving force for change.

At the turn of the millennium, after "about 35 years without a design overhaul," some technologists believed that the phone was due for some changes, many of which are already occurring. After all, they asked, "Why . . . must we peck out seven-digit codes on a keypad just to talk with someone?" In fact, the time has already come when we can dial many, if not most, phone numbers by pushing a memory, select, or redial key. With the development of voice-recognition technology, we may never have to dial a number again, and the phone keypad may go the way of the rotary dial. But for the time being, the number pad is still the basic device that it was in the 1960s.

In the keypad experiments of that era, dialing time was found to be approximately equal among the rectangular array of two horizontal rows, that of the three-by-three array with 0 below 8, and two circular arrays. Circular arrays most closely mimicked the then-familiar rotary-dial arrangement. Indeed, one circular configuration of the numbered push buttons placed them in the respective hole positions on the conventional telephone dial—that is, C-shaped, with the digits increasing from 1 to 0 in a counterclockwise direction. This arrangement was, understandably, called the "telephone" arrangement. Alternatively, the number keys were arranged in a circle open at the bottom, with the digits progressing from 1 to 0 in a clockwise fashion. This was called the "speedometer" arrangement.

A study found the difference in the five most preferred arrangements, in terms of keying time, to be insignificant. Interestingly, the now-familiar telephone-keypad arrangement, then called the "right-reading 3 by 3 plus 1," did not produce the fastest keying time, nor was it the most preferred. The most preferred arrangement was the "two horizontal rows" of buttons, and the fastest in keying time was the arrangement that mimicked the conventional rotary dial, the "telephone." The two horizontal rows of buttons were used in early trials, in which a standard telephone had its rotary dial replaced with the keypad arrangement and was mounted on a special table so that the seven-inch-deep prototype push-button mechanism could project out the bottom of the phone but still be hidden from view. This allowed the test telephone to have familiar-looking proportions. In the final analysis, however, it was the now-conventional telephone-keypad arrangement that was chosen, "since it uses the available space efficiently and permits a simplified design in the initial application." In other words, the final choice was a judgment call and a compromise, in that the keypad arrangement, after the mechanism was reduced in size, fit neatly and reasonably attractively into existing telephone bodies. It is difficult, however, to identify one single consideration that led willy-nilly to that choice. Some observers believe that it was simply the "not invented here" syndrome that led the phone company to choose over others the keypad arrangement that it did.

Regardless, the arrangement of the push-button telephone keypad was firmly established by the early to mid-1960s. Some who resist technological change or who have little motivation to go along with it still use rotary dials four decades later. A facility with a rotary dial, with its roots in the alphanumeric telephone numbers of an earlier era, does perhaps provide an advantage in dialing those numbers that we are supposed to remember by mnemonic devices, such as FOR-FOOD, which could designate the telephone number (367-3663) of the local pizza-delivery service. Such devices might even have seemed natural when we were used to telephone exchanges, but in the all-numbers world of today, they can be a nuisance. I, for one, find using my cordless phone frustratingly slow when I have to input words rather than numbers, whose places I have come to know like the back of my hand.

The old rotary telephone dial had, of course, no letters of the alphabet associated with the 1 hole and only the OPERATOR function was associated with the 0 hole. This left eight finger holes to accommodate the twenty-six letters. Including them all would have introduced an asymmetry in the design, and so the letters Q and Z were not used. This did not present an insurmountable problem, since the original function of the letter holes—to dial telephone exchanges—was easily satisfied by avoiding Q or Z as one of the first two letters of exchange names. Familiar names not only made sense because they corresponded to neighborhoods or sections of the city but also because it was probably thought to be easier to remember a couple of letters and five digits than a string of seven nondescript numbers. I can still remember the telephone number my family had when I was a teenager: LA 7-3902. (The LA stood for Laurelton, which was the community next to us, where, I assume, the actual exchange switches were located.) However, I cannot remember subsequent all-digit phone numbers we have had, except parts of them, like 1066, which, being the date of the Norman invasion of Britain, had some extratelephonic significance. Now that named telephone exchanges are history, the advent of text messaging has made it desirable to have all twenty-six letters squeezed onto the numeric keypad, and so Q and Z have been added in various places. I have one cell phone that has squeezed PQRS onto the 7 key and WXYZ onto the 9, but that solution is far from standard.

Regardless of its alphabetical quirks, I have come to like the telephone keypad mostly for its invaluable assistance to me in remembering my various "PIN numbers," as they are redundantly called. I remember these access codes not so much by their digits as by the pattern my finger traces out in keying them in. Indeed, I have become so accustomed to remembering the PIN for my ATM card by its pattern that I have to punch out the pattern in my mind if asked to recite the numbers. People with spatial memories can also remember telephone numbers easily by the pattern they trace on the keypad.

The telephone-keypad arrangement was about a decade old by the early 1970s, when the first handheld electronic calculators began appearing on the market. The Texas Instruments prototype dates from 1967, and it had not an LED window for displaying results but a thermal printer. Most important for this discussion, however, is the fact that it had a keypad not with the telephone arrangement of buttons but with the now-familiar 7,8,9 buttons in the top row. This was the natural way to arrange the calculator keys, because by this time it was the way they were arranged on most desktop adding machines, which were increasingly following a British Standard dating from 1963, which had gotten a fingerhold just as the telephone keypad was being introduced. As early as 1968, human-factors researchers recognized that "it seems highly likely that in the near future many people will be frequently involved in two numeral data-entry operations, using two keysets with identical configuration but very different numeral arrangement. Before even considering the question of confusion in concurrent use of two different layouts, it is worth asking whether one or the other is more efficient for its purpose."

All of the concerns and studies of the ergonomists were soon moot, however, for the future did come, as it always does, and with two established key sets. The companies that competed for the early calculator business did not have the luxury of time that comes with a monopoly for conducting extensive human-factors tests. Pocket scientific calculators, which date from 1972 and were immediately embraced by engineers and scientists, all came with the 7,8,9 keys in the top row. As if to emphasize the independence (or thoughtlessness) of the designers of such keypads, the placement of the 0 was another story altogether, as was what accompanied it in the bottom row of buttons. The 0 could be in any of the three positions and still is, and it could share the bottom row with the decimal point, +/–, =, π, or a host of other keys.

Few if any calculator or telephone users appear to have become finger-tied, however, and whether adding or dialing, we all seem to adapt easily to the machine before us. The fact that users of technology can switch freely between two keypads with similar geometric arrangements but with grossly different key designations illustrates not conformity to technology so much as mastery of it. There appears to be something in our makeup that allows us to adjust immediately to the task at hand. This is not an indictment of the human machine that can do this, but admiration for it.

And we do it well beyond the telephone and calculator. My son's coffee table has four or five remote-control devices sitting on it, each with a different key and button arrangement, some of which are different still from the cellular telephone and electronic calculator nearby. He uses them all as if he were a percussionist in an orchestra, who does not miss a beat in switching from block to triangle to cymbal to timpani to glockenspiel. Technology, like music, is enriched by variety.

FIFTEEN

Selective Design

GETTING IDEAS for new designs comes more easily to some than to others. Leonardo da Vinci is famous for his sketches of machines and other devices, many considered well ahead of their time. Readers have seen in his Renaissance notebooks the germ of some ideas that were not realized until the twentieth century. One bridge that he sketched was only built five hundred years later, when a footbridge based on the design was completed near Oslo, Norway. Leonardo's ideas may have taken so long to come to fruition in part because he wrote his commentaries on them in his own mirror language. Also, his notebooks did not come to public attention during his lifetime; in fact, they went largely unread for centuries.

Other inventive minds of the Renaissance were more interested in taking advantage of the new technology of the printing press and thereby communicating with their contemporaries far and wide. Some authors created what have come to be called "theaters of machines," which were essentially catalogs of inventions conceived, drawn, and disseminated, although often unbuilt. Among the most celebrated of these works is the book on diverse and ingenious machines written by the Italian military engineer Agostino Ramelli and published in 1588. Though filled with a good number of drawings and commentaries on deployable bridges and siege machines, the book also contains designs for mills and water pumps. Works in the tradition of Ramelli continued to be published throughout the seventeenth and eighteenth centuries, providing inspiration, if not actual plans, for machinists, inventors, and engineers.

In the nineteenth century, it became popular to catalog the numerous mechanisms and devices that had become part of the real and imagined mental toolbox of inventors and designers. Some of these books came to dispense with words entirely and presented just page after page of illustrations of mechanical movements and mechanisms that produced a specific output for a given input. Elementary and not-so-elementary combinations of gears, screws, ratchets, cams, toggles, cables, sliders, cranks, and the like were carefully drawn and arranged in close proximity to related devices. The inventor or designer needing a way to convert linear to rotary motion, for example, could go to the appropriate page or pages of a suitable theater of machines and pick out what he needed. If it was not there, the browser might be inspired to come up with still another design adapted from those in the catalog.

As late as the first part of the twentieth century, it was common to build physical models of such devices, thus adding a third dimension to the depictions and also providing the opportunity to turn the gears or crank the handle of a device of interest. Not only were these mechanisms attractive embodiments of the way things fit together and move, but they were instructive and inspirational as well. Some of these kinds of models survive on display in a stairwell of the Museum of Science and Industry in Chicago. Mostly, however, such models now reside only in the attics of museums or on the shelves of collectors, the World Wide Web hosting new virtual theaters of machines. That is not to say that designers are no longer grounded in reality.

Dennis Boyle is someone who has collected curious little things since he was a boy. He used to keep these odd artifacts, strange materials, and clever mechanical parts in a cardboard box, which he called his "magic box," under his desk at IDEO, where he worked as a designer. At brainstorming sessions, something from the box could provide the spark, the inspiration, the idea for something new. When Boyle became a studio leader at IDEO, he wanted to share his box of things with his coworkers, many of whom had little gadgets and mechanical trinkets of their own tucked away in a desk drawer or in a corner of their cubicle. Thus was the origin of IDEO's Tech Box, a cabinet on wheels, with large compartmentalized drawers full of technical curiosities. The contents are cataloged, along with information on sources and references, like books in a library or artifacts in a museum, and made openly available to IDEO staffers on the firm's own intranet.

The day I visited IDEO, the Tech Box was parked in what appeared to be its usual central location in the studio. To give me a flavor for the Box, Boyle opened a couple of drawers and took out two small rubber balls, some lengths of flattened copper tubing, and a few other innocuous-looking items that one might find in a five-and-dime store or at a flea market. He dropped the balls onto the cabinet top and one bounced back into his hand, while the other stuck to the surface. It was not sticky, but was made of a material that just had no bounce to it at all. The oddness of the ball's behavior qualified it for the Tech Box, for who could know when someone in the group might find inspiration in or application for a rubbery-looking material that had no bounce? The presence of the bounceless ball in the Tech Box was evidence that such a thing existed.

The copper tubing proved equally fascinating. Boyle had my host, the autodidact Scott Underwood, who had invited me to tour IDEO, pour a cup of boiling water from a kettle that was at the ready, suggesting to me that what I was being shown was no impromptu demonstration. The cold eighteen-inch-long tube was placed in the hot water, and I was asked immediately to touch its end. To my surprise, the end sitting a good foot above the water was already too hot to hold. Boyle called it "a heat pipe" and explained that sealed inside the otherwise-evacuated closed copper tube was a small amount of moisture that had become superheated steam almost instantaneously. The tube of copper, which is an excellent conductor of heat anyway, had sucked the heat out of the coffee cup as a straw hooked up to suction would the water. It is heat pipes like this, Boyle explained, that enable computers to be cooled without the annoying noise of a fan motor.

I suspect that other drawers of the Tech Box, which remained closed in my presence, held even more mysterious devices, but perhaps they were for IDEO eyes only. Like a good magician, Boyle gives away only

The Tech Box at the product-design firm IDEO contains a large variety of odd and unusual materials and gadgets, which can inspire new applications.

secrets that are already common knowledge. However, it is becoming increasingly difficult to keep magic, or any other secrets, in a closed drawer. The tradition of theaters of machines that lives on in the Tech Box is also manifesting itself in the peculiar Web sites and powerful search engines available on the public Internet via the World Wide Web.

Just as a solitary writer might open up a thesaurus to find a synonym for a word in order to convey a shade of meaning, so an independent inventor or designer can now open up a googol of Web drawers in the hope of finding something odd and unusual that might be just perfect for some new design application. The World Wide Web enables customized theaters of machines to be assembled in seconds. However, as much as a creative Internet search may produce a plethora of fascinating things to look at, selecting from among them the very one that will make a new design click is not as easy as finding the mot juste in a dictionary. Designing goes beyond language, and there are a lot more things that can be imagined, drawn, and assembled than named. Indeed, why else do we so often have to refer to a thingy, a thingamajig, a whatchamacallit, or a doohickey? If we can't name something, we have to describe it, which might take thousands of words, or somewhat fewer pictures. Designers and others have spent many a lunch hour trying to describe something new and unnamed.

St. Michael's Alley is a small restaurant in downtown Palo Alto, located just a few blocks from IDEO's buildings. It is a favorite of designers and other locals from nearby Stanford University and the surrounding Silicon Valley. As if to emphasize that it caters to a creative clientele, St. Michael's covers its tables with paper, at least at lunch, making them not unlike drafting boards. The practice reduces the laundry bill, of course, but it also invites sketching and drawing, something that certain kinds of diners are likely to do anyway. As I recall, some family restaurants even provide crayons for kids to use on table paper, an activity that their parents no doubt hope the kids will have gotten out of their systems before the good tablecloth is used at home. Table-paper drawing is encouraged in many restaurants where architects, engineers, and designers tend to congregate, as they do in St. Michael's Alley, and the day I ate there with a group of IDEO engineers and designers, they demonstrated with their pens and pencils that they were accustomed to drawing while eating. But even at times other than business lunches, restaurants can be the setting where design takes place.

Going out to eat in the evening with some friends is itself an exercise in creativity, and the common design objective is to have a pleasant and satisfying meal together. Like all design problems, a number of subproblems must necessarily also be solved along the way to realizing the final goal. The design team's leader, called the host or the organizer, is often paternalistic, if not dictatorial, selecting the restaurant himself and inviting the party of guests to it. Sometimes, however, if a reservation has not been made, the decision of where to eat is made as a group, perhaps over cocktails. Brainstorming might take the form of members of the group calling out types of restaurants or areas of town or even specific establishments.

When the group is in its hometown, the members can know a good deal about the local restaurants and even have had direct experience with some, if not all, of their culinary designs. But this does not always help, because where one person had a wonderful meal on one occasion, on another occasion someone else may have received slovenly service and an unintentionally cold entrée. On a given evening, one member of the group may want nothing but steak and another anything but steak. One may want to stay close to home and make it an early evening; another may want to combine dinner with entertainment or a floor show and close the joint. If the dinner is dutch treat, the bottom line may come into play more explicitly than otherwise. After a round or two of drinks, the only thing that may be clear is that the group will have to reach a compromise.

A classified catalog of restaurants—the Yellow Pages of a phone book—might be consulted, which can lead to new choices, places that no one has remembered or even tried. If the group feels adventurous, it may decide to try an entirely new place, to make a prototype of the meal, so to speak. With the restaurant's phone number in hand, one of the group will check on the availability of a table that is the right size (party of four) and the right style (nonsmoking). One may be readily available, in which case it is reserved for an appointed time. If there is no table available at 7:30, will the group wait until 9:00? The maître d' is put on hold while the dinner party holds a design conference. They agree to eat a bit later than planned. (Maybe they will have another drink.)

When the party arrives at the restaurant and is shown to its table, the group finds itself immediately facing another design problem, one that everyone knew they would confront but to which none has given much forethought. Nevertheless, a decision has to be made on the spot, and preferably without much discussion. Who will sit where? If the party comprises two couples, will they sit next to or across from one another? That is, will the table be boy, girl, boy, girl or boy, boy, girl, girl? The seating arrangement can easily influence the nature of the conversation and thus the ultimate design of the dining experience. At the moment when the party is approaching the table, other design choices may also become apparent. Who will sit facing the kitchen at this last vacant table in the restaurant? If the table sits people side by side, should the party quickly consider who is right-handed and who left-handed?

Once the seating compromises have been made (or dictated by the host), the quartet faces even more design decisions. When a group of diners walks into a restaurant, the individuals may or may not have a good idea of exactly what their meal will be, but merely by choosing a particular establishment, they have pretty much assured themselves of the kind of meal it will be. Neighborhood Chinese restaurants generally do not serve filet mignon, and fancy French bistros do not serve hamburgers. Regardless of the restaurant type, the diners are presented with choices, sometimes in columns headed A, B, and C and sometimes in categories such as Appetizers and Entrées. Ordering dinner in a restaurant is like designing a meal from the theater of dishes known as a menu.

Even though part of a group, the individual diner is ultimately a lone designer, and the menu is the sole supplier's catalog. The waiter is the salesman, the table is his territory, and the kitchen is the factory where the parts or components of the meal are made to order, manufactured on demand, and delivered just in time. Each individual meal is the ensemble of its courses, and the ensemble of meals for the table is the collective dinner of the group, its end product. In the process, what one person orders can affect what the others order. Some people like to eat the same thing so that they can experience the meal without sharing. More commonly, they enjoy having different things to sample as a group, passing bread and butter plates and fork- and spoonfuls of food across the table. Whatever the predilections of the diners, there comes a time when they have to stop anticipating. They must stop reading the menu and talking about it. They have to cease doing research and make some hard decisions. Then they must send their orders beyond the swinging doors and await their designer dinners.

As all diners know, each restaurant menu item, each component of the meal, has, in fact, already been predesigned by the chef out of a combination of several ingredients. In many fusion restaurants, these ingredients tend to be exotic and surprising, if not seemingly incongruous, in combination, or at least not ones normally found on the same plate. Whereas one diner may want to order the smoked duck as an appetizer, she does not like the beets that go with it. Her partner likes the steak but not the parsnips. His friend loves salmon, but not in vanilla sauce. The host prefers the halibut, although not the oysters that garnish it. In other words, unless they are willing to try to second-guess the chef's designs by asking that this ingredient be omitted or that be substituted, they must choose among a limited number of dishes, none of which may have everything they like and nothing they dislike. The social pressure of a dinner party can discourage omissions and substitutions, and so each diner/designer may have to make serious compromises. These compromises might involve ordering a second-choice appetizer because it comes without the onions that the diner's first choice includes, or, worse, selecting a third-choice entrée because the first two look heavy on the garlic.

One restaurant that we frequent varies the design problem by discarding its menu on certain nights. The kitchen has only such and such ingredients this evening, the waiter informs the diners, but they can order them fixed in any way they want. The superabundance of choice leaves many a diner tongue-tied and ordering a favorite meat or fish fixed any way the chef thinks best. Others try to stump or dumbfound the chef. In the end, the chef has the greatest fun and, perhaps because of a mischievous diner's request, may just happen on a new idea for a dish.

On the typical evening in a typical restaurant, however, the menu is fixed, and the waiter proceeds around the table, taking orders, sometimes without writing them down. After he leaves, everyone will wonder how he keeps it all straight, which he usually does. Humans can develop extraordinary powers of memory when they are engaged in an important specialized task. In any case, the last at the table to order may have found himself changing his mind and looking for alternative selections as each of his tablemates chooses something he was planning to order. He may not want to appear to be copying their design, even though it is not patented. He might be able to get away with selecting the same appetizer as the person seated across the table and the same entrée as the diner to his left, but even that might show a lack of originality. Finally, he decides to order two dishes that are entirely different from those that he had had in mind just a moment earlier, making compromise on compromise in a split second for no other reason than a psychological one.

Diners can also be greatly influenced by economic factors in designing their meals. When there is a host, everyone else hopes that she will order first to give them a clear signal as to whether to select both an appetizer and an entrée, and in what price range. Even when the meal is dutch treat, economic considerations can still influence meal design, in that most diners do not, as a matter of form, want to order the most expensive starter and entrée when others at the table are going for the least expensive. Such considerations also apply when separate checks are not available and it is understood that the bill will simply be divided by the number of diners. Still, I have been in groups where this situation was taken by some participants as license to order the highest-priced items on the menu, regardless of others' choices, showing no shame at getting a bargain at everyone else's expense.

Ordering wine with dinner presents yet another problem in design. Having both red and white with the meal is a familiar solution, seeming to obviate great compromise, but, in fact, only doubling it by requiring choices to be made among both reds and whites. Yet two bottles of wine on the table can appear to be pretentious, or at least intemperate for small parties of light drinkers. Cost is also an important consideration, of course, with wine lists often showing a wide range of prices. Frequently of late, I have found myself in a restaurant with people who prefer to order wine by the glass, so that all diners can make their own selection. Wine lists seem increasingly to encourage this by offering a good range of wines in single servings, but seldom can every wine on the list be ordered by the glass. The glass option seems a wise dinner design choice, in that everyone gets to match wine with entrée, but not drinking the same wine also makes the meal seem less communal, each diner more solitary with his or her own vintage and palate. No design choice is without its pros and cons, its constraints and compromises.

Design choices do not end with the main course. There is the inevitable dessert card to contend with, increasingly as varied and involved as the dinner menu. In some restaurants, a dessert decision is requested when the entrée is ordered, to allow time to prepare a warm soufflé or a freshly made apple tart. Some restaurants are not so considerate, and diners are informed of the extra time needed for the freshly baked chocolate cake when they receive the dessert menu. This means, of course, that anyone who wishes to have it will force a timing decision on the entire table. Even if dessert is forgone, there is still the decision about coffee—regular or decaf, espresso or none—with or without an after-dinner drink. All of these design decisions naturally have long-range effects that can have an impact not only on our digestion and sleep but also on our long-term health.

Some diner/designers solve the problem of deciding on dessert by never having it, treating it as an optional design feature that they do not opt to choose. Others, who do like to cap off a meal with dessert, want to have their cake and eat it, too. Yet when they are involved in a well-developed dinner conversation, they do not like the intrusive interruption that accompanies the reading of a dessert menu, thus turning the conversation to the menu itself, or to the presentation of the trolley display by a garrulous waiter. Some diners obviate the intrusion of dessert choices by always having the same standard thing, such as sherbet, fresh fruit, or cheese, all of which are usually available in a good kitchen. Once, on a transatlantic crossing, my wife and I were tablemates with a Scottish couple who invariably ordered cheese and biscuits for dessert. After a few evenings, the waiter did not even have to disturb our conversation, as he automatically brought what had become the standard dessert of the table. On another occasion, when dining in Dijon, my wife and I found that ordering cheeses for dessert caused a selection of several dozen to be wheeled to our table, totally diverting our talk and thoughts. There were far too many different cheeses to try. Even with the help of the knowledgeable waiter, we could order only a portion of the cheeses and so could not be sure our choices were the best. We could not know what we had missed.

After the dessert plates have been cleared, the design of a dinner is still not complete. There is the matter of the check, the bill, *l'addition*—the moment of reckoning, when it must be decided whether the meal designed and consumed truly satisfied the usually unstated but frequently abused constraint of cost. In American restaurants, the waiter often brings the check with a solicitous offer of a second cup of coffee. In France and Italy, however, the check often has to be begged for, and catching the eye of a waiter who does not want to appear anxious to dun the customer can inspire exaggerated contortions in pantomime. In Europe generally, once the bill is received, there is usually no further design decision left to make, other than perhaps what credit card to use, since the service is almost always added to the bill. Unlike the practice in American restaurants, where the waiter and your credit card disappear for a while and sometimes you get someone else's credit card and bill back, the standard procedure in Europe is for the waiter to process the credit card at tableside, producing from a wireless handheld computer terminal the printed chit to be signed. In the United States, since the tip is generally not included in the bill, the diner is faced not only with an economic design decision tinged with ethics and politics but also one that requires a numerical calculation. Many people always leave the same percentage of tip, employing their own design algorithm for establishing the amount. For a 15 percent tip, one that works well is this: Divide the bill by two, multiply by three, and move the decimal point one place to the left. This does not address the question of whether to base the tip on the total bill or on the pretax bill, but such fine points are but design details. For the diner who prefers to tip according to service, more qualitative choices have to be made.

Leaving the table can present further design decisions. One evening, my wife and I were having dinner at a local restaurant. There was an uncommonly large open space beside us because the table for four normally in that space had been moved between two others to accommodate comfortably a party of eight. The rearrangement made the normally ample aisles to the door very constricted, especially since people were sitting back from the table as they talked over coffee. In leaving, we had to make ad hoc decisions about how to design our exit.

Upon leaving a restaurant, further choices present themselves. Should one take a mint or not? Should one take one mint or two? Should one stop at the rest room or wait until one gets to the basketball game or whatever other event has been planned for the evening? Such little decisions do contribute to the overall design of the evening, of course, and trivial as some of them might seem, in the aggregate they can have an enormous effect on how we will feel about the evening.

Some restaurants—especially chain establishments—design their entire interior and their menu according to a unified theme, thus providing a prepackaged (predesigned) dining experience. Though the fare may consist of little more than variations on hamburgers and french fries, the menu can be expected to contain enough clever nominal variations on the restaurant's theme to give the illusion of more design choices than there actually are. Indeed, the predesigned meals and the milieu in which they are eaten may distract the patrons from the fact that the food is not really very good at all. Fast-food restaurants are essentially theme restaurants, though their menus do not always carry through on a theme so creatively. Rather, these restaurants are like food factories, turning out hamburgers and fries as uniform in taste and as reproducible in packaging as anything can be that is made and sold by the billions. The greatest design choice faced in such a restaurant can be whether to order a packaged meal or to supersize it.

Fast-food restaurants also often offer standardized meals for children. This kind of predesign is not unlike what moms and dads used to do in packing a lunch for a child going to school. In my grade school days, lunch boxes or brown bags typically contained some peeled and cut-up carrots or celery sticks, a sandwich that might or might not have varied with the day of the week, and a cookie or a brownie. The school often provided the beverage, which was usually a small carton of milk. If we made any design choice at all, it was between white milk and chocolate.

The idea of a predesigned meal does not necessarily diminish its quality, however. Indeed, many a young tourist has been introduced to French food through the set menus offered in the countless small restaurants on the Left Bank in Paris. These prix fixe offerings, advertised on signboards everywhere and featured prominently on restaurant blackboards, are welcome prepackaged respites from the long day of choices facing any tourist. Yet even they can present choices, such as smoked salmon or a filet steak offered for a surcharge.

Wherever we dine and whatever meal we design, menus as catalogs are part of the experience. Unless we want to go shopping and then cook at home, thereby choosing from the larger number of ingredients in the catalog of a supermarket, we are pretty much limited to what is on the menu. Since it is the rare restaurant or supermarket that completely revises its menu or shelf offerings daily, we are limited in our choices and our combinations of them. And new dishes are seldom without reference to former ones. It is similar with design, as the new thing almost always has parts and properties that we have seen and used somewhere before.

SIXTEEN

A Brush with Design

A toothbrush is one of the first things we use in the morning and one of the last we use at night. In between, the weapon of choice to keep dentists at bay once hung silently at the ready in its bathroom holder, its bristles drying for the next job. Increasingly in the late 1990s, however, toothbrush users began to lay their dental arms down in a variety of new positions. Some toothbrushes were left, like umbrellas in a stand, leaning against the rims of plastic glasses, their damp bristles curling up and out like bad hair. Some users just set their implements down on the bare countertop, placed haphazardly in a puddle beside the sink. Others, who did not like picking up a soggy brush, balanced theirs on the edge of the counter, allowing the bristles to air-dry with an assist from gravity. Less daring users set theirs down on a clean, dry washcloth, like thoughtful maids do in some hotels.

The reason for the experiments with new storage positions was simple: The latest styles of toothbrushes did not fit into the old standard holders that had for decades been built into medicine chests and ceramic-tile walls. Neither did they fit into the holders made of chromed steel that were screwed into wood and wallboard; nor were they compatible with the plastic ones fastened to wallpaper or mirrors with unforgiving double-sided adhesive pads. The toothbrushes also did not fit into the countertop decorator accessories used by homeowners who did not wish to attach things to any surface or by apartment renters who were not supposed to do this, either. Suddenly and without warning, countless holders, whose familiar round and oblong holes could not accommodate the new fat-handled toothbrushes, became obsolete.

Some social scientists might argue that it was all a plot by the manufacturers of toothbrush holders, in collusion with the toothbrush companies and the home-improvement industry, to force everyone with a bathroom to remodel or at least redecorate so bathroom accessories could handle the new-style toothbrushes. A student of technological change might speculate that the toothbrush companies focused so intently on redesigning the old handle and on coming up with a new bristle pattern that their managers, engineers, and designers simply did not think about, or care, whether the new brushes would fit into the existing infrastructure. They might as well have designed a toothbrush without worrying about whether it would fit in the average-size mouth.

Most of us use our toothbrushes the way we use our socks: encountering them while we are half awake or half asleep, and giving little thought to how they were made. We have long ago forgotten what an arduous childhood task it was to learn how socks are put on or taken off. At in-between times, we tend to forget that socks even exist. It is only when one loses a mate or develops a hole in a toe, which is bothersome more because it is uncomfortable than unsightly inside the shoe, that some attention is paid to these tubes of wool or cotton. Likewise, the toothbrush tends to come under scrutiny only when its bristles wear out or when it no longer fits into the place thought to have been made especially for it.

The standard toothbrush of the early 1990s, whether complimentary with a dental cleaning or bought at the drugstore, could indeed be a frustrating tool—and one as difficult to fit as a pair of shoes. Toothbrushes came in a variety of sizes, for children and adults; of bristle patterns, some with clever patterns of color that signaled when they should be replaced; and of stiffnesses, usually designated soft, medium, and hard, none of which ever seemed to be just right. Regardless of what kind of bristles they held, however, most of these standard toothbrushes had a standard handle. It was straight, flat, and hard, and generally of an unremarkable shape, its most distinguishing feature being that it fit like nothing else into the toothbrush holder. Different toothbrushes were distinguished mainly by the color of their plastic handles, but the colors were seldom compatible with bathroom schemes. Some siblings and roommates, after discussing what colors to buy, often got confused about which brush was theirs anyway.

The first toothbrush is believed to have taken the form of a "chew stick," which was simply a twig whose end had been masticated into a frayed state. By rubbing this brushlike device against their teeth, Egyptians millennia ago achieved a state of primitive dental hygiene. In time, abrasive pastes made of powdered pumice (an ingredient in pencil erasers) and vinegar were applied to the chew stick. As late as the twentieth century, artifactual descendants of such sticks were still used effectively in some African tribes, who fashioned them from a "toothbrush tree." In some rural areas of the United States, such implements were known as "twig brushes."

The simple toothpick, often in the makeshift form of a splinter of wood or the point of a knife, in some cultures remained the tool of choice for cleaning the teeth, as it had been in Roman times. Bone and ivory have also served as effective toothpick materials, as have silver and gold. I have a Swiss Army knife that incorporates a plastic toothpick among its many concealed devices. But the toothpick, like the twig brush, because of its short, straight construction, can be difficult to apply equally effectively to all teeth. The development of the modern toothbrush, with its bristles set at a right angle to the relatively long handle, did enable those hard-to-reach teeth to be cleaned more effectively.

Toothbrushes made of stiff hog bristles inserted into handles of bamboo or bone are said to have appeared in China about five hundred years ago. These brushes found their way to Europe, and there the hog bristles came to be replaced with the softer ones of horses. However, horsehair brushes were too soft to be effective, according to an eighteenth-century French treatise on dental surgery. Such criticism naturally led to the use of a variety of different animal hairs, none of which was found to be perfectly suited to making the perfect toothbrush.

In 1938, brushes made using the new material nylon were introduced in the United States under the brand name of Dr. West's Miracle Tuft toothbrush. Because moisture does not penetrate nylon, toothbrushes made of it dried out more effectively, leaving less chance for bacteria to grow between uses. Furthermore, whereas natural bristles tended to pull out of the brush and lodge between the teeth, the new nylon ones were fixed securely enough into the handle so as to obviate that annoying occurrence. Unfortunately, the early nylon bristles were also very stiff, and so they tended to tear the gums and cause bleeding. Chemists at Du Pont, the company that developed nylon, were put to work on the problem and eventually came up with a material that was easier on the gums. The resulting Park Avenue toothbrush, with softer bristles, could be purchased for forty-nine cents in the early 1950s, when a toothbrush with hard nylon bristles cost only ten cents. In subsequent decades, toothbrush styles proliferated.

Few of us think about the fine details of how a toothbrush is made or works, and that may be why we have to go to the dentist regularly. In fact, using a single toothbrush to clean all thirty-two teeth may be asking too much of one tool, given the different characteristics of incisors and molars, the different accessibility of the cheek and tongue sides, and the different orientation of the left and right sides of the mouth relative to the brushing hand. I used to switch-hit when playing softball, but I am not comfortable switch-brushing. I use my right hand exclusively, and so am one who must turn his shoulder, arm, hand, and brush this way and that to reach the various parts of his mouth with bristles at something approaching an appropriate angle.

What monodexterous and graceless brushers like myself probably need to brush our teeth most effectively is a suite of specialized toothbrushes, say a half dozen or so, each of which has a handle angle and bristle orientation designed to reach a different part of the mouth. This way, we could pick up and put down, like a dentist filling an improperly brushed tooth, the proper tool for the task at hand. We could use one toothbrush for the outside of our left molars, another for the inside of our right incisors, and still others for the various other categories of tooth surfaces. Such a kit of toothbrushes has not been designed, produced, or marketed, as far as I know, but it certainly might be effective in preventing dental caries—if we could remember which brush to use on which teeth. To use such a set of brushes we would also have to possess the patience of a sloth and the time of a tortoise. In the chill of the morning and the still of the night, chances are we would end up using only one of the brushes for our entire mouth. And what bathroom would have enough toothbrush holders or counter space to accommodate so many different brushes?

In the mid-1990s, the state of the art of toothbrushing was such that manufacturers saw a promising opportunity for introducing not a suite of toothbrushes but a redesigned single one. In 1997, Colgate began marketing its Wave, an appropriate name for a toothbrush with a serpentine handle, which was also thicker. Some consumers soon registered complaints about its incompatibility with standard holders, but the manufacturer "stood its ground." According to an associate director of toothbrush development for the company, "Holders shouldn't dictate your design." The constraint of fitting a brush handle into something intended to accommodate it was once "the No. 1 priority" for designers, but now it was said to be "letting the tail wag the dog" for people to "let a bathroom fixture control their oral health." Whether or not manufacturers like Colgate told designers, "Full speed ahead, and damn the holders," the holders were damned—often without explanation. According to one design critic, "Toothbrush packaging is preciously small and doesn't leave much room for lengthy explanations," for Colgate or its competitors in the race to distinguish what have been called "the athletic shoes of the late 1990s."

Oral-B Laboratories had been making toothbrushes since the 1950s and had contributed to the proliferation of specialized models. It was Oral-B that targeted the children's market in 1984 with a Star Wars character line, followed soon thereafter by a Muppets line. In 1995, Oral-B was acquired by the Gillette Company, which, like many other corporations, had diversified since its founding. It should not have been a surprising acquisition for a company that was founded on the premise that money was to be made in selling not the razor but the replacement blades, for toothbrush bristles, like blades, wear out. But rather than just new bristles, a whole new replacement brush was expected to be bought. And just as razor manufacturers have to introduce new models regularly in order to maintain their market share, so Oral-B felt that it had to introduce a new model to remain competitive. Of course, a new model involves design, or at least redesign.

As companies often do, Oral-B contracted design firms that specialize in coming up with new products, or at least new takes on old products, to supplement their own design teams. One firm that the toothbrush company involved was IDEO—whose credits at the time included such now-classic designs as that for the original mouse for the Apple computer—to look afresh at children's toothbrushes, which were "pretty much just smaller versions of adult brushes." Among the first things that the IDEO design team did was what might be called field work: Young kids were observed bushing their teeth. What the team noticed and focused on was the fact that, as opposed to older kids who held a toothbrush with their fingertips, little ones gripped it with their entire fist. The recognition of this "fist phenomenon" led to the seemingly paradoxical conclusion that toothbrushes designed for kids should have fatter handles than those intended for adult use. (This should not have been surprising, since children lack the fine motor skills of adults and cannot easily manipulate everything exactly the way their parents do. Thus, young children prefer to write and draw with larger-diameter pencils, often gripped with the whole fist.)

The toothbrush design that IDEO came up with led to Oral-B's Squish Grip line, whose fat, soft, spongy, yielding—almost fleshy—handles made them inviting and easy for little kids to hold. The Squish Grip brushes looked more like toys than like dental tools, which encouraged children to spend more time cleaning their teeth. IDEO also worked on the redesign of an adult model for Oral-B and came up with the Gripper, which was a precursor to what has come to be an infamous fat-handled toothbrush.

It took another research team three years to come up with that new design. Among the things the team discovered in the course of its work was that people brushing their teeth held their toothbrush handle in five distinct ways. These were given descriptive names: power grip, spoon grip, precision grip, distal grip, oblique grip. This last was found to be the most popular among those brushers observed in the study, but people were also found to use a combination of several of the basic grips while brushing. Not surprisingly, the hand moves about the toothbrush as it is angled and maneuvered around the teeth. Whether designing a suite of five toothbrushes, one for each of the five grips, was ever even considered is moot. As inventors invariably must do, the design team decided to go the way of compromise. Oral-B chose to develop a single unique new toothbrush that would be all things to all grips.

What resulted was described as a totally redesigned toothbrush—twenty-six patents were secured to describe all its new features. The toothbrush was named the CrossAction, referring to the pattern of its bristles, some of which are crisscrossed. This assures that at least a portion of the bristles are executing a scraping action as the brush is moved back and forth. A typical old standard brush had all the bristles projecting straight out of the brush head, which meant that they had more of a broomlike sweeping action. Anyone who has used a push broom knows that it usually leaves a lot of dust behind. Pulling the broom against the grain of the bristles is much more effective in getting a sidewalk clean. It makes sense that the same should apply to a toothbrush, as long as the brusher moves the brush in the appropriate manner.

The ultimate motion of the bristles of the CrossAction or of any toothbrush is controlled through its handle, and so its design was critical. Oral-B knew that its new product needed an ergonomically designed handle that would accommodate any grip, but the handle also had to be aesthetically pleasing in color, form, shape, and feel—and, of course, it had to be manufactured reliably and economically. Another outside consultant, the California firm of Lunar Design, was engaged, and they used computer-aided industrial design tools to help them in their task. A total of eight handle designs were tested with different focus groups from around the world, so that cultural biases would not be overlooked. The eight handle configurations were narrowed down to three, which were further tested as prototypes before a final design was settled on.

Naturally, the new toothbrush was designed to answer objections to the old. Standard toothbrushes tended to have a slender handle that was shorter than it needed to be to get the toothbrush head well into the mouth and agitate it vigorously against the teeth. The cheapest of toothbrushes had annoyingly small heads, few and too-soft bristles, and thin, flexible handles that felt like plastic forks in the hand. (On a recent visit to China, my wife and I found such toothbrushes, along with microtubes of unusual-tasting toothpaste, in every hotel room. It was convenient to have a fresh toothbrush at hand as we waited for our luggage to arrive, but the effectiveness of the brush was minimal and the stiffness of its bristles lasted little longer than the tiny tube of paste.)

The Lunar Design team wanted the CrossAction toothbrush to have a handle that fit comfortably and snugly in the hand, filling it rather than teasing it. The resulting handle is much larger than normal, and it feels good to hold. (The growing popularity at the time of the Squish Grip, the Oxo Good Grips potato peeler, and other large-handled kitchen tools made this an opportune time to introduce such a handle for a new adult toothbrush.) Part of the positive feel of the CrossAction handle has to do with its rubberlike gripping features, complete with thumb stops, which allow for more control even with a lighter grip. The handle has an indescribable irregular shape, which might nevertheless be described as funky, curving and widening and narrowing all over the place. It also has different textures and colors, reminiscent of designer running shoes. The handle on the prototypes was even thicker than that on the model eventually marketed. What caused the slimming down was the discovery that the CrossAction toothbrush, like the Colgate Wave, would not fit into standard toothbrush holders.

This potentially disastrous complication with the handle size was apparently not a problem to the Oral-B designers in their studio, but to members of the study groups of consumers on whom the toothbrush was tested. Is it possible that none of the designers was curious enough to take a prototype home and use it under everyday conditions? Or did the secrecy and security associated with new product design not allow such a natural trial? While such conditions may have prevailed, it is also possible that the designers were so focused on the function and

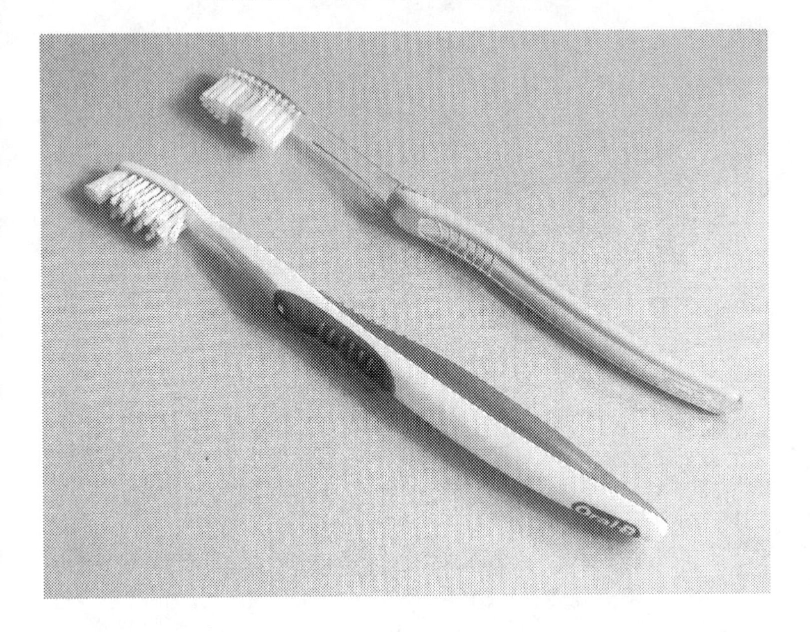

The Oral-B CrossAction toothbrush does not fit into standard holders; the redesign of the Oral-B Indicator model was constrained so that it did fit into them.

look of the toothbrush that they did not use it in context. Just as the designers of the Brita pitcher did not seem to think much beyond how it filtered water, so the designers of the CrossAction toothbrush seem not to have thought much beyond how it works in the mouth and feels in the hand. A similar shortsightedness occurred when the designers of Becky, the handicapped friend introduced into the Barbie doll family in 1997, made her wheelchair too wide to fit through the door of Barbie's dollhouse. The noncompliance with the Americans with Disabilities Act was discovered not by the toy designers or by the toymaker Mattel, but by young girls playing with the dolls. Then again, it may be that the manufacturer's is the true explanation for the Oral-B toothbrush being put on the market as it was: "The ergonomic value it was delivering definitely outweighed the fact that it couldn't fit into a hole that was designed in the 1950s."

Adults enjoyed using the CrossAction toothbrush; even if they did not know where to put it to dry, they loved the feel of it. The designers had gone back to the electronic drawing board to trim the handle's girth, but they could go only so far without losing the form and feel they had worked so hard to develop. Few design projects are without snags and setbacks, usually in multiple occurrences. When it had come time to make prototype handles, the CrossAction team found that the thick forms were warping during the injection-molding process. The warp was as much as two tenths of an inch, which is significant and noticeable in a six-inch-long handle. All of the careful, if not tedious, computer modeling to get just the right look and feel to the handle was beginning to appear as if it had been for naught. However, the problem was finally solved in a way that structural engineers have long used to account for the sag that results when heavy beams are laid between their supports. Just as steel or concrete bridge members are cambered when formed so that they will sit flat when installed, so the CrossAction toothbrush handle was given a counterwarp of two tenths of an inch in the mold so that when it cooled it would take the desired and now-familiar final shape.

As many a user knows, the CrossAction still does not fit into most toothbrush holders, but Oral-B decided to market the new product anyway. Those who bought the oversize brushes were given the opportunity to order a holder that would accommodate them. My wife and I, who were very pleased with the ergonomics of our CrossAction toothbrushes, found our own larger-holed holder in an out-of-the-way gift shop. A local pottery studio that makes ceramic toothbrush holders has Oral-B CrossActions sticking out of the holders on display, showing that the infrastructure industry has responded to the opportunity, if not to the need.

The case of the Oral-B CrossAction toothbrush represents clearly the choices and compromises (and changes) that must be made during the process of developing virtually any design. The five toothbrush grips that users may or may not employ were taken into account in designing the single handle, which could not be expected to fit perfectly for each grip or each user, let alone for each different hand that would pick up the brush. The ironic discovery that the toothbrush handle, which was designed by advanced computer technology to fit ergonomically into as many hands as possible, did not fit into existing toothbrush holders illustrates how easy it is to think only just so far in a design problem. The Oral-B team evidently did not think out of the box in which the toothbrush was to be sold. The very feature that the designers had so focused on as an asset proved at the same time to be a liability. Fortunately for Oral-B, consumers were forgiving and adaptable in trying a new thing, and so the ergonomic advantages of the new toothbrush far overshadowed its decorator disadvantages.

At the beginning of the new millennium, just about when the CrossAction was becoming familiar, Oral-B, perhaps seeing in this situation of its own making still another "product opportunity gap," decided to redesign its classic Indicator toothbrush. This severe hard-plastic-handled toothbrush, named for its rows of colored bristles that faded with use to indicate when it was time to buy a new one, had been developed in the late 1980s by a design team at Braun, the Gillette-owned German appliance maker that has been described as "über-modernist." The Indicator went on the market in 1991, and over the course of ten years, which is apparently a long life for a toothbrush model, more than 1.25 billion of this "minimalist piece of hand-held architecture that was mission-specific" were sold in America. But in 2001, sales were down, due largely to the proliferation of large-handled designs that Oral-B itself had encouraged, and Braun's design studios were given the task of updating the classic Indicator.

The design team was headed by Till Winkler, who admitted that in Germany they do not have "those crazy things" that Americans call toothbrush holders. Nevertheless, many holders that once stored the familiar Indicator toothbrushes between uses were now sitting empty throughout the United States, and so one of the principal explicit constraints that Winkler's team had to work within was that the handle of a redesigned Indicator had to fit through the oblong holes of an American toothbrush holder. Another constraint was cost, with the charge being to come up with a newly designed toothbrush that could still be sold for under two dollars.

The new Indicator, which became available in stores in 2002, has been described as a "younger, more curvaceous model." However, unlike the CrossAction and other fat-handled toothbrushes that curve in every direction, the handle of the redesigned Indicator curves in only one. From the front or back, its handle looks flat. Its slender plastic curves, caressed by the soft rubbery materials that have become de rigueur for toothbrush design, are readily apparent only when it is viewed from the side. Thus, the handle can be snaked into the holder like a woman getting into a tight dress. Where the business end of the old Indicator had rigidly straight lines, that of the new toothbrush also has curves, resulting in a much softer-looking neck and head.

Regardless of how strikingly fresh any new design might look, the oversight of something so important as compatibility with the existing world can jeopardize years of development work and result in an ultimate design and financial disaster. In many a new product, overlooking the compatibility of a design with the existing infrastructure has changed commercial promise to commercial failure even in the largest of enterprises. One gigantic oversight occurred in the 1850s. The versatile Victorian engineer Isambard Kingdom Brunel set out to design the largest steamship in the world. The *Great Eastern,* measuring 692 feet in length and weighing 32,000 tons, was so much larger than shipyards had been accustomed to dealing with that it had to be built and launched from an unconventional position—parallel to the water's edge. The launch took an embarrassing three months to complete, and thirty years later the ship ended up being cut apart for scrap. In between, it suffered as many embarrassments as triumphs. Developed to sail from England to Australia on the single load of coal it carried, the huge ship proved to be too large for most harbors and so had limited sailing options and thus limited opportunity to repay its investors. It did lay the second (successful) transatlantic cable, but in its last years afloat it also suffered the indignity of being turned into a sideshow attraction, a freak, as a gigantic floating amusement park moored to a dock in Liverpool.

In more recent times, perhaps the supreme symbol of high technology also proved to be correctly designed for all but making money. The Concorde supersonic airliner was generally agreed to be the most beautiful airplane ever built. Traveling at an altitude of 55,000 feet at a speed exceeding Mach 2, it could go from New York to London in less than half the time that the trip takes a conventional subsonic jet airliner. However, since the plane had to execute elaborate flight patterns because of the noise it emitted on takeoff and landing and because it had to avoid generating sonic booms until it was over open water, the Concorde was limited to use in only very few cities of the world. To the end of their design life, the small fleet of Concordes that were built made regularly scheduled flights only to and from New York, London, and Paris.

The design of anything, whether it be a toothbrush, an ocean liner, or a supersonic airliner, is clearly tricky. There are always challenges to meet and compromises to be made just to get the thing to work properly. During the design process, those challenges and compromises are naturally met and made in the design office and laboratory, where the design team becomes intimately familiar with the object it is creating. The thing's minor shortcomings tend to be overlooked and its idiosyncrasies come to be viewed by the designers as affectionate characteristics. After all, this is a new thing, in whose ways owners will be instructed by diagrams on the back of a package or via some type of user's manual. Other questions ancillary to primary function, such as where the thing will rest between uses, or how it will dock in ports built for smaller ships, or where it will be allowed to fly, do not appear on the radar screen. That is not to say that these are not valid design issues or serious design obstacles, for clearly, in the whole scheme of things, they are. It is just that design is hard work, and when the fancy handle is warping in the mold, or the supership is sticking on the ways, or the superfast prototype plane is becoming unbalanced on takeoff, more immediate technical design and redesign questions tend to displace the remote ones. Because designers look so closely at the thing they are designing, they can tend to see its ultimate context, like the handle of a toothbrush in use, as a great peripheral blur, if they see it at all.

SEVENTEEN

Design Hits the Wall

EACH TIME my wife and I have bought a house, we have done so amid conflict and contradiction. It was not that we argued over our decisions; it was that they always had to be made when there were competing objectives. We purchased our first house in Austin, Texas, under the self-defined, simple, but essentially incompatible constraints that we wanted to live in a nice house in a good neighborhood with a good location and good schools, but we did not want to pay more than a certain price. Obviously, we had to compromise somewhere, and we did so by spending more money than we had originally hoped to. We did get a nice house in a desirable part of town, not too far from my office, but the house had some quirky features to it and we were not left with much of a budget to alter, furnish, or decorate the place. After the initial euphoria of moving in, we realized we would not live very happily ever after in it.

Before long, we began to drive around the neighborhood, looking at houses that seemed at face value to be more like what we imagined we would like to live in. In some cases, we went with a Realtor to walk through a house and see if its space inside was as attractive to us as the house and landscaping had appeared from the street. We came close to buying a stone house of handsome proportions that had the most wonderful living room and master bedroom. The former was bathed in sunlight when we looked at the house one afternoon, and the latter was similarly illuminated when we returned to look at the place again the next morning. Unfortunately, the floor plan did not allow any space for a study for either my wife or me, there was no family room for our children, and the kitchen was small and in need of work. The stone construction would not have made it easy to expand or change the structure without ruining its wonderful unity, and we knew at the time that we could not afford to do it right. So we passed on buying the house—and in subsequent months, we sometimes drove by just to admire it from afar.

We lived in our first house for several more years, making minor decorating changes here and there, mostly in buying bedroom furniture for our growing family. We continued to keep our eyes open for houses going on the market in the neighborhood, but increasingly we knew that the expense of moving was not something we needed at the time. Even though we felt our home was less than ideal for us, we had bought location, location, location, and so we felt good about our investment in real estate. Indeed, we had little trouble selling the place when we relocated so that I could take another job. The new position paid considerably more, and so we thought we could afford to buy a better house, but in the Chicago suburban market we found little to choose from, even in our new price range. It was definitely a seller's market, and after we closed the deal on the house we settled upon, we had to live in a motel for a month until the departing family moved out. The house was nice and had adequate space, but not nearly as much as our growing children were coming to need. My wife set up her typewriter in the breakfast nook, and I put my desk in the basement. In spite of our unorthodox use of its space, we fixed up the house to our liking and lived in it comfortably and happily, but not forever after.

Within a year or so, we again began to long for a house with real office and study space and with more room for the children, and on a street with less traffic. Our kids could also have used a larger backyard. And it would have been gratifying to have a two-car garage to accommodate our automobiles and bicycles. So we started to walk and ride our bikes around the neighborhood, which we liked, watching for FOR SALE signs. Whenever we came across a house that looked to be about the right size and style, we made an appointment to inspect it. As quirky as we thought the arrangement of space had been in our house in Austin, we were not prepared for how strangely imaginative house remodelers could be. We saw one house with a screened porch accessible only through a bathroom, another with a bedroom that could be entered only through another bedroom, like a railroad flat in Brooklyn. In the end, we did not find anything in our price range that was a clear improvement over what we had and that seemed to warrant the cost of the change. And so we stayed put until I again took a new job, this time in North Carolina.

The housing market in Durham gave us more choice, but not more options. With this move, I gained in independence but not in salary, and so price was again a significant factor. We decided to try to live close to my office, arguing that if I could walk to work, we could get along with just one car.

Having already bought two houses, we were seasoned home buyers, and so we knew that the chances of our finding the perfect home were about as good as winning a door prize at a large convention. Most perfect houses, we had come to believe, were already owned by perfectly happy families who were not interested in moving or selling. Furthermore, it became clear that what may be the perfect house for one family may not be the perfect one for another. How does a family find the perfect house?

If you can't find the perfect house on the market, have it built, one real-estate agent told us. By building, you get to choose your own lot, how your house is situated on it, how it is laid out, and how it is appointed. Naturally, it will therefore be the perfect house for you and your family. On the other side of the building argument, my wife and I had heard horror stories from couples who had had new homes built for them. They had felt overwhelmed with choices—everything from shingle color to cabinet hardware—and, after moving in, they discovered that their perfect house had leaks in the roof and cracks in the walls. We were not tempted. The few small jobs that handymen had done for us confirmed that no horror story was likely to be exaggerated.

Building a house is ultimately a problem in design. Like every such problem, the process of solving it is at the same time liberating and constraining. It is liberating because you can start with a blank sheet of paper, on which you can draw anything. You can put doors and windows and rooms where you desire them, and within reason, you can make them as large or as small as you wish. But sketching in doors and windows on a floor plan is a lot different from putting them into brick and plaster walls, where an open door can block a closed window. Building is also constraining in that the choices can be very limited, unless you chose to upgrade at a price. Above all, the process is constraining because you usually have a budget, beyond which it is not wise for you to go.

In fact, there are so many design constraints and decisions to be made in building a house that most people concede most of the choices to an architect or builder, who has had the experience of having made similar ones many times before. The architect, having heard what the home-building couple wishes in their perfect house, will prepare drawings with a professional (or computer-assisted) hand. The architect will (usually) be sure the doors do not block the windows, will locate the dining room next to the kitchen, and will remember to fit a powder room in under the stairs to the second floor. For a structurally daring house, the architect will likely consult with a structural engineer about how to frame a great vaulted living room or support an unusually large balcony. As is the case with all design decisions, there will be numerous unavoidable issues of constraint and compromise.

Constraints on a design are often imposed by a third party. In the case of a house, the town or county usually has a building code, which can limit the height or size of a structure. This, in turn, limits the number of rooms that can reasonably be incorporated into the design. Not all constraints carry the force of law, but social pressures can be just as constrictive. Building an ultramodern home and painting it gaudy colors in an established neighborhood of conservative redbrick Georgian mansions would not be neighborly. While it may not be against the law, it would not be in keeping with what a reasonable person might be expected to do. Social pressure, one hopes, provides sufficient disincentive.

When designing and building a house, we are also limited by constraints associated with the nature of materials and the sizes in which they come. We may not like the appearance of joints in the woodwork, and so we might ask the contractor to order special lengths of crown molding or siding, which will naturally cost more. Even if we are able and willing to pay the premium, it may not be possible to find a supplier for extra-long pieces of cove molding for the family room or oversized tiles for the master bath. While no laws of nature may prevent the manufacture of such oversized supplies, we may not want to wait as long as it might take to have the special order filled. The nature of the constraints on what we can do in designing or building a house, or any other structure or machine, forces us to compromise.

As close as my wife and I ever got to building was the addition of a room to a house that we owned in Durham, something we did in spite of our oaths not to do so. Our adventure was a model of constraint and compromise. The house itself had not been our first choice. It was one of several that we looked at around the Durham area in our three-day house-hunting visit a few months before I was to begin my new job. Just before we returned to Chicago, we made an offer on the house that we thought was the best choice from among about a dozen that seemed likely possibilities. The house was affordable, had a nice floor plan, and was adaptable to our needs. Though it had one more bedroom than we needed, that room would serve as a study that we could share. (At the time, my wife did her writing early in the morning and I did mine late at night, and so we did not think we would get in each other's way.) It was not our absolute dream of a home, but it was the best we thought we could do under the circumstances. On the drive back to Chicago, we talked about how we would redecorate and arrange our furniture.

Shortly after we arrived home, we received a call from the Realtor, informing us that there had been another offer on the house and that it had been accepted. This left us without a place to move into, and so, by telephone, we made an offer on what we had agreed was our second choice. It was also a nice house, but very different from our first choice. Whereas we had once thought we would be moving into a traditional house with a two-car garage, we now were willing to settle for a very modern house with no garage at all. Perhaps more important, the house we did buy was smaller than we wanted, and so we lived without a room for our desks until we decided where we might build an addition.

Since the lot was also small, the west side of the house had been built almost on the property line, preventing any expansion in that direction. The north side, the front, presented a thoughtful arrangement to the street, one that we did not want to alter. Besides, the only places onto which we could add an extension in the front were to the kitchen or a bedroom, neither of which option made sense. Building in front of the kitchen would have left that room without its large picture window, which not only provided light to the space but also opened it up to the trees and the birds that frequented them. The south side of the house was its most striking asset, with a broad deck running all along its length. The living room wall was all windows, providing a wonderfully open view of our tree-filled backyard and letting in the winter sun to warm the uninsulated structure. Also on the south side was the master bedroom, whose large and only window looked out into the backyard. Any additions that we made to the house would have to be off the east side.

That side was set back a good distance from the street, to which it was parallel. Unfortunately, the utilities easement along the street meant that we could build out from the house no more than about ten feet. We also did not wish to eliminate windows on that side of the house, which left us only about three feet of wall space through which to cut a door. We decided that this would be the entrance to our future study, which would be narrow but long, stretching south into the backyard and at the same time screening the deck from the street. This happy coincidence gave us an added degree of privacy, which we welcomed. The final dimensions of the study would be about ten by twenty feet, and the entire east wall would be lined with bookshelves and cabinets, thus providing some much-needed storage space in a house without an attic and with only a partial basement. The south wall would have a large double window, and there would be a pair of sliding glass doors facing west and leading onto the deck. We were proud of our design and made sketches of it to show prospective contractors.

It was only after we had begun to talk to builders and to receive their written bids that we realized fully how much of an adventure in contradiction, constraint, choice, and compromise the construction of a simple one-room addition could be. Among the first things we learned was that what one contractor said was impossible, another said was easy.

Since our living room had a cathedral ceiling of sorts, the roof over it was naturally sloped. We had hoped that the roof over the addition could just continue that slope, presenting a seamless appearance from the street. One contractor told us that this was no problem. Another told us it was impossible, because it would make the ceiling height at the east wall as low as six feet. Not only would that cause us to lose two twenty-foot-long bookshelves, which we did not want to do, but it was also not a legal ceiling height. If we insisted on the sloped roof, he said, all he could do was build the floor two feet lower, which meant we would have to step down into the study and up onto the deck. The stairs would effectively have narrowed the room at those locations beyond what was already minimally acceptable, so we agreed to abandon the idea of a sloped roof. We would have to accept a flat roof over the study, thus introducing a break in the overall roof profile of the house. We agreed we would have to make that concession or abandon the idea of an addition altogether.

In all, we got four bids on the job, ranging from about $10,000 to around $25,000 in 1981 dollars. The lowest was from the contractor who told us it would be no problem to give us a sloped roof. After he had begun the job, he undoubtedly would have "discovered" that we would have to accept either a lower floor or a flat ceiling, and either choice would have increased the cost, since he had not figured either consideration into his bid. We agreed to go with a bid that was close to but not the lowest, because we felt most comfortable with that contractor and he told us he could begin the job immediately and be finished in about ten weeks.

Work did begin right away, but it progressed in fits and starts. The first thing to be done, of course, was to cut a trench around the periphery of the room, pour a concrete base, and build a foundation wall. This was done by a mason, who worked quickly. When he was finished, we saw for the first time in real scale what the dimensions of our room would be, and we were concerned that it was so long and narrow. There was no turning back, however, and so we convinced ourselves that it would look fine when we moved our long, narrow desks into the room.

With the foundation in place, it was time to frame in the joists to construct the floor. In order to begin that phase of the project, a doorway had to be cut through the living room wall, so that the new floor could be constructed at the same level as the old. This was perhaps the most traumatic part of the job for us, because it involved altering a part of the main house in an essentially irreversible way. With large, loud saber saws, the workmen attacked the wall with a casualness that shocked us. Having cut all the way through the drywall and the exterior cypress siding, as well as through the framing in between, the workers pushed the wall out so that it fell into what would be a new crawl space. (With the interior of the exterior wall exposed, we also confirmed what we had surmised the previous winter: There was absolutely no insulation in the wall.)

As he was trimming the edges of the door frame to remove some tenacious splinters, the builder remarked, "They don't make them this way anymore," which we first thought was just a comment on the absence of insulation or merely small talk. Then he showed us with his tape measure that the two-by-fours really were just about two inches by four inches. Our house had been built in the 1950s, when the dimensions and name of lumber were more nearly one and the same, for today a typical "two-by-four" is only one and a half inches by three and a half inches. We felt doubly sad that the workers had to saw through an endangered—nay, an extinct—species of lumber, which survived only fossilized behind Sheetrock. But the deed had been done.

When we had invited bids, we had taken the prospective contractors into the basement of the house to show them the underside of the floors. Pointing up at the subfloor's tongue-and-groove boards crossing the floor joists at a forty-five-degree angle, we indicated that we wanted the addition to be constructed in just the same way, with hardwood flooring above. The contractors had all nodded and made a note. We were thus surprised one afternoon when we found the new room's floor made up of four-by-eight-foot sheets of plywood. Where was the neat diagonal subfloor that we had specified? The builder explained that that exact kind of lumber was no longer made, and anyway, it was too costly and time-consuming to construct subfloors that way. He thought the plywood would serve just as well. We were surprised at this breach of contract, but the contractor gave us little satisfaction. He essentially told us to accept it or he would leave the job. Besides, he said, after the hardwood was installed, no one would ever know the difference. We would, of course, but we acceded, accepting the first of many differences between what we thought we had agreed upon in the specifications and what we found the contractor installing.

In all our discussions about how many windows and bookshelves we wanted in our new room, we were oblivious to other practical matters. Among the things we had never thought to consider were how to relocate the outside water spigot, which would be covered up by the addition, and how to get heat and air-conditioning into the new room. The contractor had no doubt encountered such details in previous jobs, or perhaps he had even learned the hard way by leaving an old spigot buried in the dark of a new crawl space under an addition and then being called back at the first sign of spring by an angry home owner left holding an empty hose. In our case, the contractor asked us, unbidden, whether we wanted the new spigot on the east or south wall. He didn't even ask about the ductwork, which he carried in a rather slipshod way from the furnace room through the original foundation wall into the new space. Like many other details, the floor vents would later have to be modified, when it was discovered that they would be covered by the bookcases, which were being fabricated separately by a cabinetmaker.

With the floor in place, the framing for the walls proceeded with surprising speed. In fact, it all happened while I was at work one day, and I came home to a fully framed, if roofless, study. We were pleased with the way the room was taking shape, and for a while we forgot our little differences with the contractor.

The next day, weather permitting, the roof would be put on. This was perhaps the touchiest part of the job, and the weather was important, because our existing roof had to be cut away along the edge of the house so that the framing for the new one could be tied into the old and a smooth (and watertight) transition made between the two. Early the next morning, a Friday, the contractor said it looked as if the weather would be fine through the weekend, and so he and his workers would begin the roof framing. They worked all day and into the evening, leaving us with a living room whose ceiling was open to the elements along the eastern edge. Since the temperatures were mild and no rain was forecast for the weekend, the contractor simply draped some plastic sheeting over the gap just to keep out the dust, squirrels, birds, and bugs. Naturally, a freak storm arrived Saturday afternoon, and a considerable amount of water ran down our living room wall and onto the carpet. We called the contractor, who came over quickly, but he said that it was too late to do anything then. He would take care of the damage after it had a few weeks to dry out.

We were angry and disappointed, to say the least, but work had gone so far that we chose to allow it to continue. We wanted the roof completed and the room walled in as soon as possible, and so we did not want to alienate the contractor. The workers did return on Monday morning and had the addition completely roofed over and walled in by Tuesday afternoon. The interior work proceeded in due course, and soon the room was ready to be painted. A door was installed between it and the living room, and things were beginning to look the way we had hoped they would. Even the living room wall, which had been soaking wet just a few weeks earlier, had been freshly painted, and it now appeared as if nothing detrimental had happened.

Among the last things the workers did was to backfill the trench around the foundation wall, which was supposed to be cinder blocks below grade and brick above, matching the original structure. In walking around the addition to give it one final inspection, we noticed that the cinder blocks at the base of the foundation were exposed where the ground sloped down into the backyard. Since this was on a very visible side of the house, it was unacceptable, and we asked the contractor to correct it. He suggested we just plant some shrubs in front of his mistake, but we did not think that this was a satisfactory solution. He finally agreed to chisel out the most offensive cinder blocks and fill in the hole with some leftover bricks that were still piled nearby. The patched wall bulged out noticeably and it remained a blemish, but we accepted it as the best he could do. What we wanted most at this stage was for the workers to finish and be gone, so that we could enjoy our new study without the sound of saws, hammers, and chisels.

The last step in the process was to satisfy the local building inspector, who objected to the electrical service now entering the house at the juncture between the old sloped roof and the new flat one. According to the local code, the pipe to which the power line connected could be no more than a few feet from any edge of the house. Since the electrical connection had not been moved before the room was added, it had to be relocated now, after all the carpentry and roofing work had been completed. This necessitated cutting a hole in the new roof, which we wished did not have to be done. The inspector was unyielding, however, and would not give final approval of the work until the retrofitting was done. Despite our fears, neither removing the old electrical post nor installing the new one through the roof created a leak, and in time we forgot about where the electric line entered the house.

We did get plenty of use and enjoyment out of our addition, but our experience with the contractor reconfirmed for us that we did not have the collective marital disposition to build an entire house. The next time we moved, we bought a house that had been newly renovated by the previous owner, a fact that was extremely pleasing to us because it meant that we did not have to deal with modifications, and therefore any workers, for the foreseeable future. This house served us well for a while, but with our children gone and our work patterns changing, its space and location were now less to our liking than they once had been. So we began again to look at houses, still seeking the perfect one for us.

We are still looking.

EIGHTEEN

Design Rising

Among the tourist attractions in Edinburgh is a Writers' Museum dedicated to three of Scotland's most famous authors: Robert Burns, Sir Walter Scott, and Robert Louis Stevenson. The museum is located in a house that was built in 1622 and acquired in 1719 by a Lady Stair, whose name it bears today. Lady Stair's House is aptly named, for guides tend to focus not on the displays of memorabilia associated with the writers who lived and worked in the building but on the curious construction of its stairway.

For me, the description of the stairs leading to the second-floor bedroom was certainly the most unforgettable part of our introduction to the house and its historic significance. Before being allowed to climb upstairs, our small group of tourists was given a lecture on the stairway's idiosyncrasy. Our guide explained to us that the person who built the house, one William Grey, was concerned about intruders sneaking up to the bedroom in the middle of the night and stealing from or attacking him and his wife, Geida Smith. As a form of protection, Grey had the stairway designed and constructed so that "the height of each of the main steps is uneven, making them difficult to run up and down." Anyone unfamiliar with the house could not ascend the darkened stairway without faltering on a step of an unexpectedly different height. The noise accompanying the misstep not only would awaken the owner but also would startle the intruder, thwarting his intentions and causing him to flee. The irregularity of the steps would also impede his flight back down them, if not actually cause him to fall, making it more likely that he would be caught.

Forewarned, we tourists were invited to climb the stairs and experience the irregularity ourselves. Since we weren't told exactly what the variation in the step height was, the man at the head of our expedition did trip on the way up, but, being on alert, he faltered but did not fall. Even though he had been warned of the irregularity of the steps, his strong predilection to find regularity in any stairs under his feet caused him to establish a rhythm of climbing that was soon thwarted. The rest of us, having observed how challenging the steps were, approached them with trepidation, experiencing the disruption of our own climbing rhythm but otherwise getting past the trap without incident, just as those residents who knew the secret of the stairway must have done even more quickly.

Even when they are not designed as booby traps, stairways can be tricky things. It seems that no two are exactly the same. Though they may have steps that are the same size, the number can vary from one or two to way too many. And even when the number is the same on two different stairways, the size of their (uniform) vertical and horizontal parts, known as the risers and treads, can vary according to the constraints of the space or the whims of the designer or builder. Indeed, it is a remarkable tribute to our ability to adapt to vastly different manifestations of technology that by paying close attention (which is possible in the daylight at least), we can negotiate even the most bizarrely designed and unfamiliar stairways with nary a fault.

Wouldn't it be nice if all stairways were not only regular but also had exactly the same proportions, the same dimensions for their risers and treads? While that would not make them effective burglar traps, it certainly could be expected to reduce missteps and accidents. In the United States alone each year, about a million people receive hospital treatment for falls on stairs, and about five thousand actually die. At least some of these accidents might be attributable to the oddities of stairways, if not to their downright faulty design.

The vast majority of those accidents occur while people are descending the stairs, which should not be surprising. If we fall when we are going up the stairs, we tend to fall forward, which means that we do not have so great a distance to go before we can catch ourselves without much difficulty on the geometry rising up to meet us. Going down, however, we fall onto stairs that are already receding from us, which in combination with the effects of gravity can make for a deadly pair of conditions.

Descending stairs is always a lot trickier than ascending them, especially when the steps have a bull-nose overhang, which shadows the step below and reduces its effective size. The overhang does not diminish the amount of tread that we have to step on while climbing up the stairs: We can place the ball of our foot well onto the tread, as long as we remember to bring our foot back a bit so the toes clear the lip of the next step. On the way down the stairs, however, we cannot so easily compensate. The distance from the edge of one tread's overhang to the next is known as the "going," and it clearly presents a foreshortened target compared to an entire tread. The going can be tough.

If the overhang is not too great, we might angle our foot such that it reaches back beneath the overhang and meets the riser below. I had clear evidence of this being done when I observed some workmen finishing an attic. The wood of the stairway up to the space had not yet been stained or varnished, and the otherwise-clean surface of the risers had been marked with numerous black heel scuffs from the workmen's shoes. Because work shoes are of the heavy steel-tipped variety, they are not very flexible in the toes. Hence, when descending the stairs, workmen necessarily tend to walk more flat-footed and so are likely to place their whole foot, rather than just the ball of it, on each stair. Especially if they are carrying something, they also try to get as much of their shoe onto the stair tread, and so the heel scrapes down the riser, leaving its mark. Taller people, who naturally have larger feet, also have to deal with finding enough room on the stair tread for their foot. Ascending stairs is not a problem, but descending, they either have to use only their heel or come down more or less sideways so that there is room for the foot to rest.

Stone steps tend not to have overhangs, and so their treads and goings are of one and the same dimension. Though all steps get worn down with age, outdoor stone steps generally show the wear much more than the indoor wooden variety. This paradox, that hard stone wears down more than softer wood, is easily resolved when we recognize that wooden steps used outdoors often tend to rot away and need replacement long before they can show significant wear. Also, stone steps outdoors are constantly subjected to the grinding action of hard shoes working sand and other grit into them. Wooden indoor stairs, often carpeted, seldom see the same tough shoes and gritty dirt.

In any case, the familiar wear marks on the stone steps of our institutions are quite telling about how we use stairs. The typical stone stairway is used very much the same way by all those who climb it, and so the wear on it is as clear as a well-hiked trail through the woods. Narrow stairways leave little room for walking anywhere but in the middle, and so the wear is very much concentrated there. Wider stairways allow for more waver by the walkers, but the wear patterns still tend to reveal the preferred paths of climbers. Americans are inclined to stay to the right on stairways and cut inside corners consistently close. Those who drive on the left side of the road should normally prefer that side of the stairs. Traffic patterns are necessarily complicated when cultures meet in cosmopolitan cities and international tourist attractions, and these patterns are all captured in the wear of the stone.

Cultural differences aside, stone stairs tend to wear out in two distinct ways: The top of the tread becomes dished out, and the once-square edge becomes rounded or, if initially rounded, cupped. The former wear pattern results from the ascent, where the shoe is scuffed forward when the foot first hits the stair and then is scuffed backward when the foot pushes off to proceed to the next step. Over time, such a reciprocal action is as effective as that of grinding a lens. On the way down, the bottom of the shoe tends to scrape across the edge of one step en route to the one on which it will rest and then repeats the scraping action on the next step down. On a narrow stairway, the wear pattern has all the characteristics of a statistical normal distribution. On a wide-enough stairway, these wear patterns follow two distinct paths, because the two-way traffic naturally segregates itself according to the dominant local or tourist patterns.

Regardless of how steps will be used or worn, they first have to be designed and built. Vitruvius, writing two millennia ago about the state of the art of building in Greece and imperial Rome, "set down the earliest known design guideline for stair layout." According to Vitruvius, who was writing about a temple:

> The steps in front must be arranged so that there shall always be an odd number of them; for thus the right foot, with which one mounts the first step, will also be the first to reach the level of the temple itself. The rise of such steps should, I think, be limited to not more than ten nor less than nine inches; for then the ascent will not be difficult. The treads of the steps ought to be made not less than a foot and a half, and not more than two feet deep. If there are to be steps running all round the temple, they should be built of the same size.

Though Vitruvius's prescription of leading with the right foot and having an odd number of steps reveals his reverence for ceremony, his tentative specification of step dimensions emphasizes the judgmental nature of design. Matters have not changed much in the twenty centuries since Vitruvius wrote, but modern domestic stairways at least are smaller in terms of risers and treads than his specifications for temples, thus making them easier to climb than ceremonial stairways.

There are actually practical as well as ceremonial reasons for having an odd number of stairs, for then it can always be the same preferred and presumably surer foot that makes the transition from floor to stair and from stair to floor. Often I find myself altering my step as I reach an unfamiliar stairway, so that I will mount it with my surer foot, my left, which, unlike my right, has never suffered a stress fracture.

People think more about stairs and their number and use than they might care to admit. How often do we overhear those descending from a tourist attraction, like the Statue of Liberty or the Spanish Steps in Rome, remark on how many stairs they had to climb to reach the top? Indeed, so many visitors must ask that many tourist locations have posted at the bottom of a stairway the number of steps that must be climbed, along with the height and other vital statistics of the given monument. The characteristics of steps and stairs are of importance to people with arthritis or a heart condition, but most of us appreciate a good set of stairs, even if we cannot quite put our toe on what makes them good.

All preconditions being equal for the design and construction of a new stairway, the dimensions of risers and goings will typically follow formulas, either by craft tradition or as suggested or mandated by standards or a local building code. (Still, two millennia after Vitruvius wrote, strict regulatory requirements typically apply only to public rather than private residential stairways.) However, all things are seldom equal, and stairways must usually be designed and built to fit preexisting conditions. This constraint can take the form of the vertical distance between the sidewalk and the entrance to a building, which in locations prone to settlement, like Mexico City, can change over time, or the height from one floor to another, which is usually a more stable measurement.

If houses were built around stairways, these could all have the same number of steps with risers and treads of exactly the same size. But stairways tend to be fitted into the frame of a house, which defines the constraints and forces the choices and hence the compromises. Imagine a two-story house framed with exactly nine feet between the first floor and the second. The number of steps in the stairway can be determined by dividing the total height to be climbed by the height of a single stair—that is, by the size of its riser. The normal range for risers today, as opposed to in ancient times, is between six and eight inches, with seven being a good mean. But dividing nine feet (108 inches) by seven inches per step results in 15.43 steps, an impossible number. The natural thing for the designer to do is to install a stairway with fifteen or sixteen steps of equal size, which will result in a riser of either 7.2 inches or 6.75 inches. In either case, the small deviation from what might be considered the ideal of seven inches will hardly be noticed.

Even if the designer does not have a Vitruvian predilection for an odd number of steps, the smaller number (with the larger risers) is more likely to be chosen, not because it is an odd number but because it involves building one fewer step. This choice is especially likely to be made for smaller buildings, because the number of steps also affects how much space horizontally (about nine or ten inches per step) must be available to accommodate the stairway. One handbook prescribed that the dimensions of the "tread and the height of one riser should add up to seventeen or eighteen inches." Another way to save horizontal space is to make the stairway steeper than normal. This is achieved by using shorter treads, which results in the stairs being more precarious. A stairway that is too steep can cause the person climbing it to feel like he is using a stepladder.

In tight interior spaces, where a straight flight of stairs cannot be fit because there would be no room for a landing in line with them, the stairway may be broken up into two runs at right angles to each other, with a landing between. Alternately, a series of wedge-shaped stairs can be introduced to turn the corner. These variations in stairway construction necessarily involve their own design constraints, choices, and compromises. Thus, the turning stairway must include steps with treads that increase in size across the width of the stair, leaving very little foot room on the inside of the turn, where most people would prefer to walk because it is the shortest distance between two levels.

I once watched a carpenter and a pair of helpers tackle the problem of building a turning stair, a construction feat that was apparently one not familiar to them. Their goal seemed to be to keep the risers uniform and all the treads as large as possible. They made a determination of how high the risers should be, but somehow they miscalculated the number of wedge-shaped steps and so cut them too small. After much discussion, they tore out their first attempt at the stairway and began anew. The second time around, they got the winding part of the stairway correct, if a bit chunky, but they somehow miscalculated on the risers of the short straight part that was to continue onto a brick landing leading to more steps going down into a basement. The last step they built had a noticeably shorter riser than all the rest, but they declared it acceptable among themselves and pronounced the job complete. We noticed the short riser the first time we went down the stairs, the forward foot coming down hard on the brick landing. Likewise, on climbing the new stairs, we tended to trip on the second step, because it was higher than the first. But before very long, like the occupants of Lady Stair's house, we had become accustomed to the anomalies of the stairway and we abandoned the thought of getting the carpenter to come back and regularize the stairs.

Whatever stair design is implemented in the construction of a home, its residents soon become familiar with the idiosyncrasies. They come to know instinctively how high to raise their foot when climbing and how far to extend it when descending the stairs. Though few of us are likely to be conscious of counting the number of steps between floors in our house, we develop a sixth sense about when we have reached the top or bottom of a stairway, even in the dark. We have neither to keep track (consciously at least) of the steps nor to look at our feet to know when we have reached a landing. Thus, the residents of Lady Stair's house must certainly have developed a familiarity with their irregular stairway, perhaps even skipping over the odd step as we might pass over a creaky one at home.

That's not to say that they may never have had an accident, for we all tend to misstep now and then. Why this happens may be more a question of psychology than of design. People have accidents when their senses are altered, such as by medication or liquor, or when their minds are so preoccupied with other things, such as a headache, that they have no room in them even for unconscious stair counting. We also have accidents when we are distracted and move in a direction contrary to good practice, such as making a sudden turn midway up the stairs because we remember something we forgot downstairs. When we do misstep on stairways, we can often catch ourselves before a serious accident results. Catching ourselves before we make a serious misstep in design is not always so easy or immediate.

Because designing a stairway or anything else involves satisfying constraints, making choices, and accepting compromises, it should come as no surprise that no stairway is exactly perfect. But just as we can adapt to stairways of every number and combination of step, to every irregularity of rise and run and creak, to virtually every pattern of wear and tear of stone and carpet, so we can and do adapt to the most imperfect of things, stairs included. Every day, we use more or less effectively a whole host of far-from-perfect tools, equipment, devices, gadgets, implements, utensils, machines, appliances, apparatuses, contraptions, and systems. Some of these are so clever and effective that they become icons of design and objects of admiration.

There are many museums dedicated to the world of things, and the Museum of Science and Industry in Chicago is one of the oldest and most widely known. Its famous neoclassical building, complete with monumental entrance columns and caryatids, is a 1930s reconstruction in limestone of the plaster Palace of Fine Arts built for the 1893 World's Columbian Exposition, Chicago's first great world's fair. Though the exterior appearance of the old building was retained, the interior of the new museum was to be done in the streamlined architecture of the 1930s.

The Museum of Science and Industry opened in 1933, the year of Chicago's second world's fair, whose theme was "A Century of Progress." The ceremonial staircase, the tall Ionic columns, and the heavy high doors were in place, but the interior accessible to visitors extended little beyond the doors. Progress on the museum had fallen short of the finishing touches; floors were still bare concrete. When the colder weather came, a primitive heating system was introduced, consisting of "metal barrels with smoldering coal inside." In time, the interior was finished, of course, but near the end of the twentieth century, the classic design problem of getting people from outside to inside the gargantuan building returned to haunt it.

Today, the scale of the museum remains enormous, though that may no longer be so immediately appreciated, for the approach to it has been altered. Most visitors now arrive through a new entrance, created in conjunction with the construction of a multilevel underground garage, which solved the museum's perennial parking problem. The solution, however, came at an aesthetic and experiential price exacted by design under constraint. Now, instead of experiencing the awe and openness associated with climbing a wide set of ceremonial stairs topped by towering fluted columns, visitors are more likely to have a sense of banality and claustrophobia as they drive their car down a long concrete ramp into an underground parking structure. After descending more nondescript concrete ramps until they find an available parking space, visitors finally can emerge from their car, but only to walk among a multitude of other cars parked beneath the low ceiling. On the main (underground) entrance level, the only real option is to walk toward a set of doors no grander than the side exit from a movie theater. This is the motorist's substitute for what was once one of the most welcoming entrances to any museum in the world. The new experience is similar to but nowhere near as edifying as descending into the museum's famous reconstructed coal mine.

Ordinary visitors who do not enter through the garage are now expected to do so through one of two new "main entrance" kiosks on the oblong circular drive in front of the building. These guardhouselike structures stand far forward of and flank the grand exterior staircase to the museum, as if defending it and discouraging its use. Upon entering a kiosk, visitors immediately have to reorient themselves, for to get to the building above, they have to descend into a modernistic cavern below, as if into a mine. To get up to the building that loomed above them from the street, visitors are directed down into a large but hard-edged and boxy space that contains the security checkpoints, information booths, ticket counters, and gift shop now associated with the entrances to virtually all large tourist attractions. Visitors to the altered space enter by way of what is, essentially, one of the museum's basements. This may work at the Louvre, but it does not at the Museum of Science and Industry.

After purchasing their tickets, visitors enter the museum proper by going up a long, narrow escalator, which is, of course, nothing but moving stairs with wide treads and variable risers that undergo the endless change of a conveyor belt. This escalator actually carries visitors *under* the grand staircase that is so prominent a feature of the building's facade, though the ceiling gives little hint of this. At the top of the escalator, where the moving stairs morph into the stone-solid floor, traffic is channeled through what is in fact an opening in the original building's foundation wall, appropriately substantial to support the great structure, although oppressive in this new context. Ironically, in a museum whose tradition has been to show the inner workings of both the natural and the man-made universe (with a giant walk-through cutaway model of a human heart remaining one of the great attractions), visitors are given no clue as to what they are passing through or where they are in the historic structure. The flow of traffic continues through a nondescript space, whose clear focus is another set of escalators, which will carry the visitors to the museum's main level, where finally they can experience the great rotunda. Here, today's museumgoers can recapture the grandeur of the space that welcomed and awed visitors in the original 1893 building. For all the architectural infelicities of its modern subterranean approach, the building retains at its core the graceful and uplifting presence promised by its grand entrance steps.

Although the interior of the dome remained unfinished and unadorned during the 1933 world's fair, its base was eventually inscribed in the gold lettering that visitors now must still crane their necks and rotate their bodies to read: "Science discerns the laws of nature. Industry applies them to the needs of man." Composed by museum curators, who were evidently not poets, the significance of the two sentences remains masked by their prosaic awkwardness. In contrast, the motto of the "Century of Progress" fair, which had the clipped sound and urgency of a telegraphic message, was decidedly catchier: "Science Finds—Industry Applies—Man Conforms."

Composition and rhetoric aside, there is a difference between the role that humans play in the two word diagrams of the triangle of science, technology, and society. In the "Century of Progress" version, man is molded by the world of science and industry, a view articulated in the nineteenth century by the essayist and poet Ralph Waldo Emerson, who wrote, "Things are in the saddle, / And ride mankind." In Emerson's time, when railroads were still young, factories were altering not only the landscape but also, for so many people, the actual way of American life. It may indeed have seemed that men and women, as well as boys and girls, were being ridden like horses, the bit of wages controlled by the machinery that was being driven not at the natural pace of the current in the old mill stream but increasingly by the high-pressured power and celerity of steam. Teams of workers, no more individuals than the clutch of horses that pulled the wagon or turned the mill, were harnessed to the machines of industry.

Emerson's view of technology was in sharp contrast to the view espoused by the engineers of the Industrial Revolution. A definition of their profession submitted in Britain by the Institution of Civil Engineers as part of its application for a royal charter, which was granted in 1828, read in part that "engineering is the art of directing the great sources of power in nature for the use and convenience of man." It is this sentiment, rather than Emerson's, that is echoed around the dome of the Museum of Science and Industry. Technology is not viewed as riding or driving mankind. Rather, it is mankind that is in the seat of a wagon, directing where it should go. It was in the telegraphic compression of the message in the 1933 fair's motto that the medium diminished the meaning.

In fact, neither version of the Chicago motto fully represents the true relationship among science, technology, and people. Each version begins with the false premise that scientific knowledge is primary. But industry, invention, ingenuity, engineering, and design do not merely apply the laws of nature, though they must work within the constraints that the laws of nature impose on all things and beings, including men and women. Often there is not even a science to apply. As one design text has put it, "Engineers use science to solve their problems *if* the science is available. But available or not, the problem must be solved, and whatever form the solution takes under those conditions is called engineering." The sources and products of that engineering, imperfect as they are, are admirable achievements. Fire was used for ages before it was understood chemically. The steam engine was "perfected" long before there was an engineering science of thermodynamics. Orville and Wilbur Wright flew before aerodynamics fully explained their feat. And bridges carried aqueducts over broad valleys for millennia before there was a modern theory of structures.

A bridge may be said to force us to conform to its technology and cross a river at the place where the bridge is located. But we can also cross the river without being conformists. We can take a boat, swim, ride a hot-air balloon, tunnel under, or maybe even wait for the river to freeze over. There are always alternatives to using a given or available technology, but when they are less convenient, less quick, or less safe, they may not be the wiser way. Not to take a bridge at its face value, as something made for our convenience and safety, is to distort the object of technology. Of course a toll bridge may also have been constructed as a financial investment, whereby those who put up the capital expect to reap a profit. And the engineer who designed and built the bridge may have had egotistical motives. Local politicians may have influenced the location of the bridge to satisfy powerful real-estate interests, including their own. Such venal motives, however, do not invalidate the bridge as a safe and convenient way to cross a wide, deep, dangerous river, or require those who wish to get from one side to the other to avoid it.

Certainly, the engineers who design and build a bridge rely on expert knowledge and experience and even science. But designing and building a piece of technology is more than an application of science. In fact, relying on science alone would make it virtually impossible to design even a modest bridge. What science would be applied? The laws of mechanics tell us that forces must balance if the bridge is to stand. But what forces, and stand how? Unless inventors and engineers, designers all, can first visualize some specific kind of bridge in their mind's eye, they have nothing to which to apply the laws of science. The creation of a bridge or any other artifact requires, before anything else, something imagined. Whether or not science can be applied to that mental construct is a matter of availability. If there is a body of scientific knowledge that can be applied, then it would be foolish not to exploit it. However, if there is none, it does not mean that the thing cannot be designed, made, and used safely. A medical treatment need not have a known scientific basis to be efficacious. There is no science absolutely needed to build a stairway. Nor is one necessary to design a toothbrush, or a drinking glass, or a meal.

In fact, "Science Finds—Industry Applies—Man Conforms" will never be more than a catchy motto. The reality is: "People Design—Industry Makes—Science Describes." It is the creative urge that drives the human endeavor of design, which leads to inventions, gadgets, machines, structures, systems, theories, technologies, and sciences. Both science and technology are themselves artifacts of human thought and effort. To ignore this in forming facile phrases that distort as they flow easily from the lips misrepresents how things really come to be and are. All made things are products of design.

Among those activists who at different times and in different places have been known as Luddites, antitechnologists, or radical environmentalists, the idea has prevailed that much technology, especially new technology, has negative consequences for society. Machines displace workers, factories pollute the atmosphere, and technological advances roll over a simpler way of life. According to the naysayers, people are not served by technology so much as enslaved, harnessed, and ridden by it. As evidence, they present arguments that computers run our lives, that we are nothing but numbers, that mass manufactures have driven out the good old stuff.

It is true, of course, that technology does sometimes seem to be working against rather than with and for us. But this is because technology, being something created and designed by people for people, is not perfect. It can never be perfect. That should not be a reason to reject it out of hand, or to fight it at every turn. Power failures may create awful inconvenience and discomfort, and sometimes even life-threatening situations. Should we therefore eschew electricity, reject refrigerators, abandon incandescent lighting? We expect bulbs to blow out occasionally, so why not power plants? Should we never expect to trip on a set of stairs? Should we cease to use them when we do? If no one can build a perfect set of stairs, should no one try to build any?

Simply put, all technology is as imperfect as its creators, and we can expect that it always will be. As we can, by practice and discipline, improve our own behavior, so we can, by experience and process, improve the behavior of our creations. In the meantime, we demonstrate our dominion over things by using and enjoying them in spite of their shortcomings. When we know where the bad stair is, it does not trip us up. As this book has suggested, there are countless examples of technology's imperfections and limitations, from the simplest to the most complex of made things. By understanding their flaws and the limitations of the design process that created them, we can better appreciate why they are and must be imperfect. All things designed and made have had to conform to constraints, have had to involve choice among competing constraints, and thus have had to involve compromise among the choices. By understanding this about the nature of design, from which all made things come, we can better negotiate the variety of stairways that we encounter, no matter how idiosyncratic or metaphorical, taking us from one level of technology to another.

ACKNOWLEDGMENTS AND NOTES, BIBLIOGRAPHY, ILLUSTRATION CREDITS, AND INDEX

ACKNOWLEDGMENTS AND NOTES

Once again, I have benefited greatly from the enormous resources of the Duke University Library System and the patience and persistence of its staff. Eric Smith and Linda Martinez were especially helpful in identifying and providing patents and other documents. I am also grateful to Diane Himmler and Tara Bowens for their help in securing a variety of materials.

My son, Stephen Petroski, collaborated with me in developing the MileAisle and SuperCirclemart concepts described in chapter 7. A patent is pending for the latter invention. My wife, Catherine Petroski, has once again been my first reader and this time has been my principal photographer. She has also been a constant source of interesting and relevant items and stories from a wide range of publications.

As always, I am indebted to the many inventors, designers, and readers who give me feedback on my essays, books, and lectures. Where I can attribute an idea in this book to a specific communication, I have acknowledged or referenced it in the appropriate place below. Full references to works cited here by the author's last name, and short title where necessary, are given in the bibliography. Where a series of quotations is taken from the same source, generally only the first or the most prominent occurrence is noted. Unless otherwise indicated, World Wide Web sites were accessed during the summer of 2002.

ONE: THE NATURE OF DESIGN

I am grateful to Fred Leibowitz, Scott Milberg, and Brian Strauss of Royal Paper Products for providing information about and samples of their company's pizza savers.

3 radio talk show: *Talk of the Nation—Science Friday,* hosted by Ira Flatow, with guests Michael Graves, Donald Norman, and the author, broadcast February 1, 2002, on National Public Radio.

5 a different public radio show: *The State of Things,* hosted by Mary Hartnett, broadcast April 17, 2002, on WUNC, a North Carolina affiliate station of National Public Radio.

5 "thingy": Cf. "Of Widgets and Whatchamacallits," *IEEE Spectrum,* September 1990, pp. 16, 18.

5 another listener, an artist: Private communication via E-mail from Nell Steelman Whitlock, July 10, 2002.

6 Another fan of the plastic tripods: See http://www.gsu.edu/~biojdsx/fowl/egg2.htm.

7 "pizza saver": See http://www.royalpaper.com.

7 "keep the top of the pizza box": "The Royal Line," Royal Paper Products catalog, p. 19.

7 less expensive polystyrene: Private E-mail communication from Scott Milberg, Royal Paper Products, September 23, 2002.

8 "good or satisfactory solutions": Simon, p. 64.

8 "decision maker has a choice": Quoted at http://www.scf.usc.edu/~mehta/readings/sciences.html.

9 dictionary definition: The dictionaries most used during the writing of this book were *The New Oxford American Dictionary* and *Merriam-Webster's Collegiate Dictionary,* tenth edition.

9 "perfect storm": See Junger.

10 "design under constraint": Wulf; see also Wulf and Fisher.

10 Sydney Opera House: See, e.g., Hawkes, pp. 98–103.

12 Robert Frost: "Mending Wall," in Frost, p. 47.

14 Donald Norman: See, e.g., Norman, *Psychology.*

14 "things would be useful": Donald Norman, on the radio talk show *Talk of the Nation—Science Friday,* as referenced previously.

TWO: LOOKING AT DESIGN

17 M. C. Escher: See Escher.

17 crystal spheres: See Escher, plate 51; see also Smith.

19 invented (first designed) millennia ago: See, e.g., Amato, p. 31.

19 A recent visit to the store: Summer 2002, Dansk Outlet Store, Freeport, Maine.

19 Bubble glass: Made by Dansk International Designs, Ltd., Belgium.

27 "as soon write free verse": Quoted in Bartlett, p. 625.

THREE: DESIGN, DESIGN EVERYWHERE

I am grateful to Diane Windham Shaw, Special Collections librarian and college archivist, David Bishop Skillman Library, Lafayette College, for providing copies of items from the library's Hugh Moore Dixie Cup Company Collection.

29 Philadelphia put filters into use: See, e.g., Steel, p. 2 and fig. 1-1.

29 Boston investors saw money: Voss-Hubbard, pp. 84–85; see also http://ww2.lafayette.edu/~library/special/dixie/company.html.

29 Machines that made paper bags: See chapter 8, "Design out of a Paper Bag."

29 Lawrence W. Luellen: See http://ww2.lafayette.edu/~library/special/dixie/company.html.

30 "Cup": U.S. Patent No. 1,032,557; see also Voss-Hubbard, p. 85.

31 raised bottom: See http://ww2.lafayette.edu/~library/special/dixie/samples.html.

31 "Cup Dispensing Device": U.S. Patent No. 1,043,854.

32 "Dispensing Apparatus": U.S. Patent No. 1,081,508.

32 Eugene H. Taylor: Voss-Hubbard, p. 85.

32 Luellen Cup and Water Vendor: Ibid.

32 "ancestor of the modern": Ibid.

32 "delivers to each and every person": Ibid.; see also American Water Supply Company, p. [2].

32 American Water Supply Company: Voss-Hubbard, p. 85.

32 Hugh Moore: Ibid., p. 84; see also http://ww2.lafayette.edu/~library/special/dixie/bio.html.

32 *Quaff Nature's Nectar:* American Water Supply Company.

32 Anti-Saloon League: Ibid.; see also Panati, pp. 122–23.

33 "Thousands of persons": Quoted in American Water Supply Company.

34 Dr. Alvin Davison: Voss-Hubbard, pp. 86–87.

34 "Death in School Drinking Cups": See American Water Supply Company, p. [5].

34 "It was broken": Quoted in ibid.

35 Dr. Samuel J. Crumbine: Voss-Hubbard, p. 87.

35 Public Cup Vendor Company: Ibid., p. 89.
35 Lackawanna Railroad: Ibid., pp. 89–90.
35 "stone coal": Finch, p. 28.
35 Phoebe Snow: Voss-Hubbard, pp. 89–90.
35 "*Phoebe dear*": Quoted in ibid., p. 89.
35 "*On railroad trips*": Ibid., in illustration, p. 90.
36 King Gillette: See, e.g., McKibben, chapter 1.
36 Health Kup: Voss-Hubbard, p. 91.
36 cone-shaped cup: Ibid., p. 97.
36 the word *Dixie:* See Bartlett, p. 475, n. 1.
36 Dixie cup: Voss-Hubbard, p. 91; see also Panati, pp. 123–24; and ww2.lafayette.edu/~library/special/dixie/company.html.
36 Dixie-Vortex Company: Voss-Hubbard, p. 97.
37 ice-cream cup: Ibid., pp. 93–96; see also Panati, pp. 123–24.
37 One brand of yogurt: Colombo.
37 plastic spoon of clever design: U.S. Patent Nos. 6,003,710 and 6,116,450.
38 A bottle of water: See Day.
41 Brita pitcher: See http://www.brita.com/402b.html.
41 call-in radio show: *Talk of the Nation—Science Friday,* broadcast February 1, 2002, on National Public Radio.
43 A competing water-filtering pitcher: The Pur Plus, available in Wal-Mart stores in the summer of 2002.
43 "Discard this water": Quoted from the box of the Brita pitcher's replacement filter.

FOUR: ILLUMINATING DESIGN

50 Fairlane Crown Victoria Skyliner: See https://app.consumerguide.com/pss/images/CAPhotoFeature.pdf.
50 "the liberating feel" and "a noble, albeit flawed, attempt": Ibid.
50 Fairlane 500 Skyliner: See http://www.musclecarclub.com/musclecars/ford-fairlane/ford-fairlane-history-1.shtml.
50 "complicated and somewhat trouble prone": Ibid.
50 automatic transmission: See Kilborn.
51 HID lights: See Paul; see also http://www.findarticles.com/cf_dls/m1068/5_57/84212313/p1/article.jhtml?term=HID+lights.
51 automatic leveling devices: See ibid.; see also Sharke, p. 71.
51 headlights in Europe: Sharke, p. 72.

54 different regulatory standards: Ibid., p. 73.
54 Maglite: Morin.
55 "set their sights" and subsequent quotes regarding Maglica suit: Ibid.

FIVE: DRIVEN BY DESIGN

I am grateful to Jan Hult, who was kind enough to seek information from Volvo dealers and designers in Sweden about that carmaker's cup holders; Catherine Petroski spoke to Volvo dealers in the United States. The author bears sole responsibility for the description of the design process.

58 first widely available true cup holders: See Haas; see also http://www.bergen.com/special/autos/cup0611.htm.
59 1997 Chevrolet Venture: Ibid.
59 "allows it to expand": Farnum.
65 CD-ROM drawer: See http://www.elsop.com/wrc/humor/cuphold.htm.

SIX: DESIGN IN A BOX

68 "of convenient size": Cooper-Hewitt Museum, p. 8.
68 "not more than twelve inches square": Ibid., p. 13.
68 "about 175,000 models": Ibid., p. 19. (U.S. Patent No. 100,000 was issued on February 22, 1870, and U.S. Patent No. 200,000 on February 5, 1878.)
68 A fire in 1877: Ibid.
68 law requiring models: Ibid., p. 8.
69 "packed into thousands": Ibid., p. 19.
69 "continuous quadrangular gallery": Ibid., p. 18.
70 microwave oven: Brown, pp. 80–83.
70 "could make a working tube": Quoted in ibid., p. 81.
70 "huge and primitive": Ibid., p. 83.
70 microwave ovens were outselling: Ibid.
70 on sale for $37.95: Seen in a Wal-Mart store in Brunswick, Maine, in August 2002.
70 the engineers Jack Kilby: See, e.g., http://www.dotpoint.com/xnumber/kilby3.htm.
71 first desktop electronic calculator: Markoff.
71 fit into his shirt pocket: Ibid.; see also Kraemer, p. 20.

71 shirt pocket–size HP-35: Markoff; see also http://www.hpmuseum.org/hp35.htm.
71 printing calculators: Kraemer, p. 20.
72 ink-jet printing: See Kraemer.
72 Ichiro Endo: "Tiny Bubbles," *Technology Review,* January/February 2001, p. 136.
73 "Since the pen's design": Kraemer, p. 22.
73 ThinkJet printer: Ibid., p. 25.
74 CD jewel box: Cf. Steve Martin.
78 notebook computer at which I am now working: Toshiba Portege 7100 series.

SEVEN: LABYRINTHINE DESIGN

80 patent attorney: *New York Times,* May 13, 2002, "Patents" column.
80 "method of swinging": U.S. Patent No. 6,368,227.
80 One-click ordering: U.S. Patent No. 5,960,411.
80 "method and system": Ibid.
81 broad interpretation: See, e.g., Steven Frank.
81 designing a shopping experience: See, e.g., U.S. Patent No. 1,720,917.
83 circular store with a rotating floor: Patent pending.
86 "a common sense approach": Swearingen.
87 "the traveling salesman problem": See, e.g., Simon, pp. 64–65.
87 What line to choose: Cf. Baker, *The Mezzanine,* pp. 117–18; see also Parker.
89 to check themselves out: See Lake.
91 Golden Gate Bridge: Golden Gate Bridge, p. 64.
92 collecting tolls remotely: See, e.g., Wade.
93 per-month service charge: Peterson.
93 eighty miles per hour: "Warnings Go Out to E-ZPass Speeders," *New York Times,* July 11, 2001.
93 The speeding problem: See "E-Z Grousing on Speed Limit in Toll Plazas," *New York Times,* July 14, 2001, pp. B1, B5.
95 As of late 2000: Wade.
95 Florida alone had five: Ibid.
95 eliminate tollbooths entirely: See, e.g., Arnold.
96 "ultraviolet laser emitter": See *New York Times,* May 27, 2002, "Patents" column.
96 "ensures that the photograph": U.S. Patent No. 6,351,208.

97 thirteen-digit ones: See Murphy.

97 crucial to solving crimes: See Gaudette.

97 The technology for intrusive surveillance: See, e.g., Markoff and Schwartz.

EIGHT: DESIGN OUT OF A PAPER BAG

I am grateful to Aydin Kadaster, who as a student doing research for me at Duke University in the late 1990s compiled much useful information on plastic bags, razors, and inventions in general.

99 The first paper bags manufactured commercially: http://ohioline.osu.edu/ed-fact/0133.html.

99 "Machine for Making Bags of Paper": U.S. Patent No. 9,355.

99 "pieces of paper of suitable length": Ibid.

100 "Machine for Making Paper Bags": U.S. Patent No. 12,982.

100 patented a third machine: U.S. Patent No. 20,838.

100 "for preventing the loss of the strips": Ibid.

100 "Paper Bag & Envelope Mach.": U.S. Patent No. 12,982, drawing sheets.

100 The way Wolle's machines formed: See U.S. Patent No. 9,355, figs. 8, 9; see also U.S. Patent No. 20,838, figs. 5, 6, 7.

101 attributed to Luther Childs Crowell: See, e.g., Krugman; see also van Dulken, pp. 180–81.

101 "Improvement in Paper-Bags": U.S. Patent No. 123,811.

101 "aware that paper-bags" and subsequent quotes referring to Crowell patent: Ibid.

101 Margaret E. Knight: See Macdonald, pp. 50–56; see also Stanley, pp. 521–25; and History and Heritage Committee, p. 202.

101 "a jack-knife, a gimlet": Knight, as quoted in Macdonald, p. 51.

102 a loom-shuttle restraining device: Macdonald, p. 51; see also History and Heritage Committee, p. 202; for alternative descriptions, see Stanley, pp. 521–22.

102 Columbia Paper Bag Company: Macdonald, p. 52.

102 "she could not possibly understand": http://www.inventorsmuseum.com/MargaretKnight.htm.

102 "drawings, paper patterns": Macdonald, p. 54.

102 her very first patent: U.S. Patent No. 109,224.

103 her second patent: U.S. Patent No. 116,842.

103 "industrial origami": I am indebted to Jim McGill of Seattle, Washington, for this term, which he uses to describe everything from folding a letter to fit into an envelope, to folding a newspaper for delivery, to folding the open end of a paper bag to form a carrying handle.

103 "to be the first to invent": U.S. Patent No. 116,842.

104 Her financial arrangements with Eastern: Macdonald, pp. 55–56; see also Stanley, p. 522.

104 Another of her patents: U.S. Patent No. 220,925.

105 shoe-sole cutting machine: See Stanley, pp. 476, 524.

105 improvements in automobile engines: See U.S. Patent No. 716,903.

106 accordion-pleated bag: U.S. Patent No. 123,811.

106 "the most simple and practical": Ibid.

109 Kraft paper: see http://www-afandpa.org/products/paperbagcouncil/facts.cfm, accessed February 19, 2002.

110 four out of five grocery bags: See http://www.plasticbag.com/environ mental/history.html.

110 5 percent of grocery bags: Amidon, p. 24.

110 60 percent of the market: Ibid.

110 Plastic bags begin as long seamless tubes: See, e.g., U.S. Patent No. 5,335,788.

110 polyethylene: See, e.g., U.S. Patent No. 5,213,145.

111 "T-shirt" bags: See, e.g., ibid.

111 recycling them has never been widely promoted: But see Amidon, p. 31.

111 over ten thousand supermarkets: Ibid. p. 24.

112 one out of every four: Ibid., p. 30, table 1.

112 "average consumer uses 500": See "Why Cloth Bags?" at http://www.clothbag.com/Our_History/our_history.html.

112 wire rack: For a discussion of some types, see U.S. Patent No. 5,213,145.

114 Sylvan N. Goldman: See Garrison, pp. 88–89.

114 His first shopping cart: See U.S. Patent No. 2,155,896.

114 In a subsequent patent: U.S. Patent No. 2,196,914.

114 the redesign of the shopping cart: "The Deep Dive: Five Days at IDEO," *Nightline,* ABC Television, broadcast in February 1999; see also Kelley, pp. 8–14.

114 Walter H. Deubner: See Garrison, pp. 86–87.

115 Robert Mentken: See *New York Times,* November 29, 1999, "Patents" column.

116 "handle-bags": See http://www.paperbag.org/seal.htm.

116 Paper Bag Council: Ibid.

NINE: DOMESTIC DESIGN

121 mow the grass of baseball fields: Leland.

123 "garlic cloves": Such a recipe also appeared on a box of Ronzoni thin spaghetti found in a supermarket in Bath, Maine, in the summer of 2002.

125 Roy J. Plunkett: See http://www.dupont.com/teflon/newsroom/his tory.html.

125 *freon:* Ibid.; see also http://users.efn1.com/~paradox/teflon/plunkett.html.

125 John Rebok: Ibid.; cf. http://www.rochester.infi.net/~rwhend/teflon.html.

126 Teflon: See, e.g., http://users.efn1.com/~paradox/teflon/plunkett.html.

126 Marc Gregoire: See http://www.uselessknowledge.com/explain/teflon.shtml.

126 Thomas Hardie: See Panati, pp. 105–6.

127 sold out within two days: Ibid.

127 Since the material is inert: See http://www.discovery.com/area/skin nyon/skinnyon970606/skinny1.html.

TEN: FOLK DESIGN

129 "right and wrong": Toobin, p. 54.

129 "There are only two things": Ibid.

129 "All we need is WD-40": http://www.illuminated.co.uk/inwo/show.cgi?card=wd40.

130 "anyone carrying duct tape": Ward.

131 "military tape": http://www.octanecreative.com/ducttape/dthistory.html.

131 "nameless, military-green product": "The History of Duck (and Duct) Tape," Manco, Inc., undated news release.

131 "durable, waterproof": Ibid.

131 Permacell: Ibid.

131 "duck cloth tape": "Duck Tape Facts," Manco, Inc., undated news release.

132 to cover up the gun ports on airplane wings: Wilson, p. 11.

132 natural thing for GIs to bring back to the States: See, e.g., http://www.octanecreative.com/ducttape/dthistory.html.

132 Jack Kahl: See http://www.manco.com/aboutus/mancoduck/default.asp.

133 renamed Manco's product: Ibid.

133 Garrison Keillor: See McGuire.

133 Duct Tape Guys: See, e.g., [Minneapolis] *Star Tribune,* August 15, 2000, p. 1E.

133 Web site: http://www.octanecreative.com/ducttape.

133 "could probably fix": *National Post,* September 16, 2000, p. N2.

133 "ultimate power tool": See http://www.octanecreative.com/ducttape/duckvsduct.html.

133 Since duct tape does not come with instructions: See "Web Surfers Stuck on Duct Tape Web Site," *Business Wire,* June 1, 2000; see also www.ducktapeclub.com.

134 NASA is said to have a policy: See, e.g., Barnard.

134 *Apollo 13* mission: See, e.g., Schlager, pp. 591–97.

135 "successful failure": See, e.g., James A. Lovell, "Houston, We've Had a Problem," *NASA History, Apollo Expeditions to the Moon,* at http://www.hq.nasa.gov/office/pao/History/SP-350/ch-13-1.html.

135 "Velcro and plain": Quoted in "Duck Tape Facts," Manco, Inc., undated news release.

136 pipe bombs: See Conlon.

136 kidnapped Philadelphia girl: See Richard Jones.

137 "certain heat-resistant standards": See "Duct Tape 101" at http://www.octanecreative.com/ducttape/DT101/index.html, accessed July 24, 2001.

137 WD-40 had its origins in war: See www.wd40.com/AboutUs/our_history.html.

138 WD-40 Company: See, e.g., Bannist.

138 DANGER: WD-40 eight-ounce spray can, Item No. 10008, purchased in the summer of 2002.

138 introducing a pair of notches: Ibid.

ELEVEN: KITCHEN-SINK DESIGN

Jonathan Cagan was kind enough to send me the galleys of his and Craig Vogel's book, which brought the story of Sam Farber to my attention. The section of this chapter on the Oxo Good Grips vegetable peeler first appeared, in somewhat different form, in *American Scientist,* November–December 2001.

141 Alfred M. Moen: Honan; see also http://www.moen.com/Consumer/about/aboutalmoen.efm.

141 make water faucets more user-friendly: Cf. Norman, *Psychology,* pp. 166–72.

141 common outlet: For a 1920s design, see http://www.pricepfister.com/website/asp/Pf_npp_pricepfisterhistory.asp#1910.

149 Sam Farber: See Cagan and Vogel, p. 14; see also Palmeri; and http://www.cdf.org/cdf/atissue/vol2_1/kitchen/kitchen.html.

149 Simon W. Farber: See http://www.farber.com.

149 Louis Farber: See Palmeri.

149 "sleek-looking steel": Ibid.

149 "best known for its colorful": http://www.cdf.org/cd/atissue/vol2_1/kitchen/kitchen.html.

149 "missed the social contact": Palmeri.

150 "product opportunity gap": Cagan and Vogel, pp. 9, 14.

150 "I heard a lot": http://www.cdf.org/cd/atissue/vol2_1/kitchen/kitchen.html.

150 Smart Design: Ibid.

150 "to come up with tools": Ibid.

150 "The design team talked": Ibid.

151 Santoprene: See, e.g., ibid.; see also Cagan and Vogel, p. 16. I am grateful to Robert J. Sauer of Granville, Ohio, for pointing out to me that Santoprene is a patented and trademarked material manufactured by Advanced Elastomer Systems of Akron, Ohio. See http://www.santoprene.com/home.html.

152 "fingerprint softspots": Ibid.

152 manufacturer was found in Japan: Cagan and Vogel, p. 16.

152 Oxo International: http://www.cdf.org/cdf/atissue/vol2_1/kitchen/kitchen.html.

152 "doesn't stand for anything": Ibid.

153 "won both customer approval and critical acclaim": Ibid.

153 new manufacturer in Taiwan: See Cagan and Vogel, pp. 16–17.

153 "knock off their own product": http://www.cdf.org/cdf/atissue/vol2_1/kitchen/kitchen.html.

TWELVE: OFF-THE-SHELF DESIGN

Portions of the material on office chairs and on the sizes of people in this chapter first appeared in *American Scientist,* November–December 2001. Other portions on office chairs appeared under the title "What's in a Chair?" in *Gallup Management Journal,* Summer 2001. I am grateful to Charley Vranian of Herman Miller for providing information on that company, and to Linda Baron for supplying images of the Aeron chair. Bill Keller furnished information on Steelcase, and Jeanine Hill sent images of chairs from the 1950s.

157 "First, it must pass the truck-driver test": Douglas Martin.
157 Charles Dickens: See, e.g., Dale and Weaver, pp. 4–5.
157 Victorian expanse of cubbyholes: See, e.g., Norman, *Things,* p. 155.
158 Metal Office Furniture Company: See http://www.steelcase.com.
158 Victor fireproof steel wastebasket: For this and other Steelcase milestones, see links on http://www.steelcase.com.
158 S. C. Johnson & Sons Administration Building: See Fiell and Fiell, *Industrial Design,* pp. 492, 494.
158 Metal Office made the steel table: See "Milestones" at www.steelcase.com.
160 Aeron: See, e.g., Fiell et al., *1000 Chairs,* p. 625.
161 Herman Miller's Web site: http://www.hermanmiller.com.
162 Galen Cranz: Cranz, p. 164.
163 *The Measure of Man and Woman:* Tilley and Dreyfuss Associates.
163 The normal distribution of heights: Ibid., drawing 11.
167 A 1960s study of chair use: See Cranz, p. 160.

THIRTEEN: FAMILIAR DESIGN

169 Doorknobs give no hint: See Norman, *Psychology,* pp. 87–92.
170 "Open the box": http://www.jmlock.com/doorknob_install-ns4.html.
170 Another set of instructions: http://www.hgtv.com/HGTV/project/0,1158,BDRE-project_7690,00.html.
171 light switch: See Norman, *Psychology,* pp. 92–99.
173 Thomas Edison: See Israel, pp. 327–28.
175 Donald Norman: Norman, *Psychology.*

FOURTEEN: DESIGN BY THE NUMBERS

The material in this chapter appeared first, in a slightly different form, in *American Scientist,* July–August 2000. I am grateful to Jeon Nam Kil, a Korean inventor, who, after reading *Invention by Design* in translation, sent me copies of an article and a standard about push-button telephones that led me to much additional literature on human factors. He also sent me a prototype of a telephone with a modified keypad—with the 0 key on the top, where he believes it is more conveniently located to input international access codes and Korean cellular and beeper codes that frequently begin with 01. He has applied for a patent for his new key-set arrangement.

180 Chesley Bonestell: See Ron Miller.
180 real space stations: See "Artificial Gravity," *New York Times,* June 4, 2002, p. F2.
181 In an earlier book: Petroski, *Invention by Design.*
182 The most commonly given answer: See, e.g., Feldman, pp. 14–15.
183 a rotary-dial telephone: See, e.g., Haltman.
183 "automatic telephone exchange": See "No Operator, Please," *Technology Review,* January/February 2000, p. 104; see also U.S. Patent No. 447,918.
183 American Telephone & Telegraph: See ibid.; for a description of using a rotary-dial candlestick-model telephone dating from 1923, see Haltman.
183 "one 10-year-old boy": Greenman.
184 Human factors: See, e.g., McCormick.
185 "few systematic studies": Lutz and Chapanis, p. 314.
186 "find out where people say": Ibid.
187 first cellular telephone call: Oehmke.
187 "most obvious finding": Lutz and Chapanis, p. 317.
187 "most other calculators": Ibid.
187 "the number of possible key arrangements": Deininger, "Human Factors Engineering Studies," p. 996.
188 field trials: "Experimental Set Using Push Buttons Placed on Trial," *Bell Laboratories Record,* August 1959, p. 314.
188 "about 35 years": Jon Fort, "Phone Keypad May Go Way of Rotary Dial," at http://www.mercurycenter.com/svtech/news/indepth/docs/moto092500.htm.
188 "Why . . . must we peck out": Ibid.
188 Circular arrays: See, e.g., Wikell, p. 36.
189 "since it uses the available space": "Experimental Set Using Push Buttons Placed on Trial," *Bell Laboratories Record,* August 1959, p. 314.
189 "not invented here": E.g., Paul Fleming, Jr., in unpublished portion of undated [2000] letter to the editors, *American Scientist.*
191 British Standard: British Standards Institution.
191 "it seems highly likely": Conrad and Hull, p. 165.

FIFTEEN: SELECTIVE DESIGN

I am grateful to Scott Underwood for inviting me to visit IDEO while I was in Palo Alto to give a lecture at Stanford University. Dennis Boyle was kind enough to show me around some studios and to introduce me to the Tech Box.

193 built five hundred years later: *Civil Engineering,* January 2002, p. 22; see also *Civil Engineering,* November 1998, p. 20.
193 "theaters of machines": Ferguson, p. 115.
194 catalog the numerous mechanisms: See, e.g., ibid., pp. 118–19, 125.
194 Museum of Science and Industry: The models were still there when I visited the museum in September 2001.
194 IDEO: See Kelley.
195 Tech Box: See ibid., pp. 142–45; see also McGrane.
195 The day I visited IDEO: November 16, 2001.
197 thingy, a thingamajig: Cf. "Of Widgets and Whatchamacallits," *IEEE Spectrum,* September 1990, pp. 16, 18.
200 One restaurant that we frequent: Robinhood Free Meetinghouse, Georgetown, Maine.

SIXTEEN: A BRUSH WITH DESIGN

208 "chew stick": Panati, p. 208.
208 "toothbrush tree": Ibid.
208 "twig brushes": Ibid.
208 stiff hog bristles: Ibid.
208 horsehair brushes: Ibid., p. 209.
208 French treatise: Pierre Fauchard, *La Chirurgien Dentiste,* as cited in Panati, p. 209.
209 Dr. West's Miracle Tuft toothbrush: Panati, p. 209.
209 Chemists at Du Pont: Ibid.
209 Park Avenue toothbrush: Ibid., p. 210.
210 "stood its ground": Maremont.
210 "Holders shouldn't dictate": Quoted in ibid.
210 "Toothbrush packaging is preciously small": Vanderbilt.
210 Oral-B Laboratories: See http://www.oralb.com/aboutus/history.asp.
211 "pretty much just smaller versions": Kelley, pp. 34–35.
211 It took another research team: "Input/Output: Not Like Pulling

Teeth," *Mechanical Engineering,* March 2001, p. 60; see also http://www.aliaswaavefront.com/en/WhatWeDo/studio/see/oralb.shtml.

211 five distinct ways: Ibid.

212 twenty-six patents: Ibid.

212 Lunar Design: Ibid.

214 Brita pitcher: See chapter 3, "Design, Design Everywhere."

214 Barbie doll: Gilje; see also Glascock.

214 "The ergonomic value": Quoted in Vanderbilt.

216 Indicator toothbrush: Pearlman.

216 "über-modernist": By Pearlman.

216 "those crazy things": In ibid.

217 "younger, more curvaceous model": Ibid.

217 *Great Eastern:* See, e.g., Petroski, *Remaking the World,* pp. 126–45.

217 Concorde supersonic airliner: For a brief critical review, see Muschamp.

SEVENTEEN: DESIGN HITS THE WALL

221 Building a house: See Kidder for a book-length treatment of the subject.

226 "two-by-four": See, e.g., ibid., p. 123.

EIGHTEEN: DESIGN RISING

I am grateful to Mark Hayward, Director of Artifacts and Archives at the Museum of Science and Industry, for inviting me behind the scenes at that institution.

230 Writers' Museum: See http://www.rampantscotland.com/blvisitwriters.htm.

230 Lady Stair's House: Ibid.

231 risers and treads: See Templer, *The Staircase: History and Theories,* pp. 173–75.

231 about a million people receive hospital treatment: Templer, *The Staircase: Studies of Hazards,* p. 4.

232 the "going": Templer, *The Staircase: History and Theories,* p. 174.

234 "set down the earliest known design guideline": Templer, *The Staircase: Studies of Hazards,* p. 26.

234 "The steps in front": Virtuvius, p. 88.
235 strict regulatory requirements typically apply: See *Structural Engineer,* November 2002, p. 11.
235 Mexico City: See, e.g., Dillon.
236 "tread and the height of one riser": Kidder, p. 267.
238 Museum of Science and Industry: See Pridmore, *Inventive Genius;* see also Pridmore, *Museum of Science and Industry.*
238 Progress on the museum: Pridmore, *Inventive Genius,* p. 52.
238 "metal barrels": Ibid.
239 reconstructed coal mine: Ibid., pp. 54–56.
240 "Science discerns": Petroski, *Remaking the World,* p. 60.
240 "Things are in the saddle": Ralph Waldo Emerson, "Ode Inscribed to W. H. Channing," quoted in Bartlett, 13th edition, p. 503.
241 "engineering is the art": J. G. Watson, p. 9.
241 "Engineers use science": Quoted in Lewalski, p. 3.

BIBLIOGRAPHY

Amato, Ivan. *Stuff: The Materials the World Is Made Of.* New York: Basic Books, 1997.

American Water Supply Company. *Quaff Nature's Nectar from* This *Chalice.* Brochure. Paterson, N.J.: American Water Supply Co., [1908].

Amidon, Arthur. "Plastic Grocery Sack Recycling." *Resource Recycling,* November 1990, pp. 24–31.

Antonelli, Paola, ed. *Work Spheres: Design and Contemporary Work Styles.* New York: Museum of Modern Art, 2001.

Arnold, Wayne. "Relief for Rush Hour: Pay As You Go." *New York Times,* August 9, 2001, p. F6.

Baker, Nicholson. *Double Fold: Libraries and the Assault on Paper.* New York: Random House, 2001.

———. *The Mezzanine: A Novel.* New York: Vintage Books, 1990.

Bannist, Nicholas. "Popular Lubricant Finds Place in History." *Milwaukee Journal Sentinel,* May 2, 1999, p. 3.

Barnard, Linda. "Eat Your Heart Out Martha. Oh the Things You Can Do with Duct Tape." *Toronto Sun,* March 20, 1998, p. 85.

Bartlett, John. *Familiar Quotations.* 16th edition. Justin Kaplan, general editor. Boston: Little, Brown, 1992.

Bayley, Stephen, ed. *The Conran Directory of Design.* New York: Villard Books, 1985.

Berg, Jim, Tim Nyberg, and Tony Dierckins. *The Jumbo Duct Tape Book.* New York: Workman, 2000.

Boggs, Robert N. "Rogues' Gallery of 'Aggravating Products.' " *Design News,* October 22, 1990, pp. 130–33.

Boxer, Sarah. "A Postmodernist of the 1600's Is Back in Fashion." *New York Times,* May 25, 2002, p. B7.

British Standards Institution. *Specification for Adding Machines. B.S. 1909 : 1963.* London: British Standards Institution, 1963.

Brown, David E. *Inventing Modern America: From the Microwave to the Mouse.* Cambridge, Mass.: MIT Press, 2001.

Business Week editors. *100 Years of Innovation: A Photographic Journey. Business Week,* Summer 1999.

Butz, Richard. *How to Carve Wood: A Book of Projects and Techniques.* Newtown, Conn.: Taunton Press, 1984.

Cagan, Jonathan, and Craig M. Vogel. *Creating Breakthrough Products: Innovation from Product Planning to Program Approval.* Upper Saddle River, N.J.: Financial Times / Prentice-Hall, 2001.

Clymer, Adam. "Tracking Bay Area Traffic Creates Concern for Privacy." *New York Times,* August 26, 2002, p. A11.

Coleman, Joseph. "Funeral Held for Pachinko Machines." [Brunswick, Me.] *Times-Record,* August 9, 2001, p. 15.

Conlon, Michael. "Mailbox Bomber May Have Run out of Duct Tape." *National Post,* May 7, 2002, p. A13.

Conrad, R., and A. J. Hull. "The Preferred Layout for Numerical Data-Entry Keysets." *Ergonomics* 11 (1968): 165–73.

Cooper-Hewitt Museum. *American Enterprise: Nineteenth-Century Patent Models.* New York: Cooper-Hewitt Museum, 1984.

Cooper Union. *A Better Mousetrap: Patents & the Process of Invention.* Exhibition catalog. New York: The Cooper Union for the Advancement of Science and Art, 1991.

Cranz, Galen. *The Chair: Rethinking Culture, Body, and Design.* New York: Norton, 1998.

Dale, Rodney, and Rebecca Weaver. *Machines in the Office.* New York: Oxford University Press, 1993.

Dallaire, Gene. "Pocket Electronic Calculators Zoom." *Civil Engineering,* February 1975, pp. 39–43.

Day, Sherri. "Bottled Water Is Still Pure, But It's Not Simple." *New York Times,* August 3, 2002, pp. A1, C14.

Deininger, R. L. "Desirable Push-Button Characteristics." *IRE Transactions on Human Factors in Electronics,* May 1960, pp. 24–30.

———. "Human Factors Engineering Studies of the Design and Use of Pushbutton Telephone Sets." *Bell System Technical Journal* 39 (1960): 995–1012.

Desloge, Rick. "Metaphase Design Has a Handle on Brushing Teeth."

St. Louis Business Journal, April 19, 1999, http://stlouis.bcentral.com/stlouis/stories/1999/04/19/focus2.html.

Dillon, Sam. "Capital's Downfall Caused by Drinking . . . of Water." *New York Times,* January 29, 1998, p. 4.

Drewry, Richard D., Jr. "What Man Devised That He Might See" [a history of eyeglasses], http://www.eye.utmem.edu/history/glass.html, accessed August 14, 2001.

Dunbar, Michael. "Locking Tapers." *Early American Life,* June 1989, p. 70.

Edison Lamp Works. *Pictorial History of the Edison Lamp.* Harrison, N.J.: Edison Lamp Works of General Electric Company, n.d.

Escher, M. C. *The Graphic Work of M. C. Escher.* New York: Ballantine Books, 1971.

Farnum, Greg. "Cupholders, Conveyors, and Cooperation." *Design News,* October 1, 2001, pp. 30–31.

Feldman, David. *Why Do Clocks Run Clockwise? And Other Imponderables: Mysteries of Everyday Life Explained.* New York: Harper & Row, 1988.

Ferguson, Eugene S. *Engineering and the Mind's Eye.* Cambridge, Mass.: MIT Press, 1992.

Fiell, Charlotte, and Peter Fiell. *Industrial Design A–Z.* New York: Taschen, 2000.

Fiell, Charlotte, Peter Fiell, Simone Philippi, and Susanne Uppenbrock, eds. *1000 Chairs.* New York: Taschen, 1997.

[Finch, J. K.] *Early Columbia Engineers: An Appreciation.* New York: Columbia University Press, 1929.

Frank, Michael. "Will Wonders Never Cease?" *New York Times,* June 21, 2002, p. E38.

Frank, Steven J. "Patent Absurdity." *IEEE Spectrum.* August 2002, pp. 48–49.

French, Howard W. "Hypothesis: Science Gap. Cause: Japan's Ways." *New York Times,* August 7, 2001, p. A6.

Frost, Robert. *Complete Poems.* New York: Holt, Rinehart and Winston, 1949.

Garrison, Webb. *Why Didn't I Think of That?: From Alarm Clocks to Zippers.* Englewood Cliffs, N.J.: Prentice-Hall, 1977.

Gaudette, Karen. "Privacy Worries Follow New Traffic Monitoring System." [Brunswick, Me.] *Times-Record,* August 12, 2002, p. 10.

Gilje, Shelby. "Barbie's Friend Finds Doors Closed." *Seattle Times,* June 7, 1997, local news.

Glascock, Ned. "Breaking Down Barbie's Barriers." [Raleigh, N.C.] *News and Observer,* February 10, 1998, p. B1.

Golden Gate Bridge, Highway and Transportation District. *Highlights, Facts & Figures of the Golden Gate Bridge, Highway and Transportation District.*

San Francisco: Golden Gate Bridge, Highway and Transportation District, 1994.

Greenman, Catherine. "When Dials Were Round and Clicks Were Plentiful." *New York Times,* October 7, 1999, p. G9.

Haas, Al. "Lowly Cup Holder Changes an Industry," *Bergen Record,* June 11, 1996, http://www.bergen.com/special/autos/cup0611.htm.

Haltman, Kenneth. "Reaching Out to Touch Someone? Reflections of a 1923 Candlestick Telephone." *Technology in Society* 12 (1990): 333–54.

Hanson, Thomas F. *Engineering Creativity.* Newhall, Calif.: published by the author, 1987.

Hawkes, Nigel. *Structures: The Way Things Are Built.* New York: Macmillan, 1990.

Hendel, Richard. *On Book Design.* New Haven, Conn.: Yale University Press, 1998.

History and Heritage Committee, sponsor. *Mechanical Engineers in America Born Prior to 1861: A Biographical Dictionary.* New York: American Society of Mechanical Engineers, 1980.

Hitt, Jack. "The Theory of Supermarkets." *New York Times Magazine,* March 10, 1996, pp. 56–61, 94, 98.

Homer-Dixon, Thomas. *The Ingenuity Gap.* New York: Alfred A. Knopf, 2000.

Honan, William H. "Alfred Moen, Whose Hands Found Need for a New Faucet, Dies at 84." *New York Times,* April 21, 2001, p. A13.

Hooton, Earnest A. *A Survey in Seating.* 1945. Reprint. Westport, Conn.: Greenwood Press, 1970.

Hounshell, David A., and John Kenly Smith, Jr. *Science and Corporate Strategy: Du Pont R&D, 1902–1980.* Cambridge: Cambridge University Press, 1988.

Hubbell, Sue. *Shrinking the Cat: Genetic Engineering Before We Knew About Genes.* Boston: Houghton Mifflin, 2001.

Israel, Paul. *Edison: A Life of Invention.* New York: Wiley, 1998.

Jewell, Elizabeth J., and Frank Abate. *The New Oxford American Dictionary.* New York: Oxford University Press, 2001.

Jones, David E. H. *The Further Inventions of Daedalus.* Oxford: Oxford University Press, 1999.

Jones, Douglas C., and Nancy Cela Jones. *Edison and His Invention Factory: A Photo Essay.* [Bar Harbor, Me.]: Eastern National Park and Monument Association, 1989.

Jones, Richard Lezin. "Officers Praise 7-Year-Old's Courage in Escape." *New York Times,* July 25, 2002, p. A10.

Junger, Sebastian. *The Perfect Storm: A True Story of Men Against the Sea.* New York: Norton, 1997.

Kanbar, Maurice. *Secrets from an Inventor's Notebook.* San Francisco: Council Oak Books, 2001.

Kelley, Tom. *The Art of Innovation: Lessons in Creativity from IDEO, America's Leading Design Firm.* New York: Doubleday, 2001.

Kidder, Tracy. *House.* Boston: Houghton Mifflin, 1985.

Kilborn, Peter T. "Dad, What's a Clutch? Well, at One Time . . ." *New York Times,* May 28, 2001, pp. A1, A8.

Kleiger, Estelle Fox. "The Better Bag." *American Heritage of Invention and Technology,* Winter 2001, p. 64.

Kraemer, Thomas. "Printing Enters the Jet Age." *American Heritage of Invention and Technology,* Spring 2001, pp. 18–27.

Krugman, Paul. "Technology Makes Us Richer: The Paper-Bag Revolution." *New York Times Magazine,* September 28, 1997, pp. 52–53.

Lake, Matt. "The Self-Checkout: Lots of Swiping, No Stealing." *New York Times,* June 6, 2002, p. G7.

Leland, John. "From Outfields to Art, One Blade at a Time." *New York Times,* July 19, 2001, pp. F1, F4.

Lewalski, Zdzislaw Marian. *Product Esthetics: An Interpretation for Designers.* Carson City, Nev.: Design and Development Engineers Press, 1988.

Lohr, Steve. "Sit Down and Read This (No, Not in That Chair!)." *New York Times,* July 7, 1992, pp. C1, C4.

Lupton, Ellen. *Mechanical Brides: Women and Machines from Home to Office.* New York: Cooper-Hewitt Museum; Princeton Architectural Press, 1993.

Lutz, Mary Champion, and Alphonse Chapanis. "Expected Locations of Digits and Letters on Ten-Button Keysets." *Journal of Applied Psychology* 39 (1955): 314–17.

Macdonald, Anne L. *Feminine Ingenuity: Women and Invention in America.* New York: Ballantine Books, 1992.

MacLean, Karon E., and Jayne B. Roderick. "Smart Tangible Displays in the Everyday World: A Haptic Door Knob." *Proceedings of the IEEE/ASME International Conference on Advanced Intelligent Mechatronics,* Atlanta, September 1999.

Maremont, Mark. "Many People Bristle at the Trend Toward Fatter Toothbrush Grips." *Wall Street Journal,* March 24, 2000.

Markoff, John. "William Hewlett Dies at 87; a Pioneer of Silicon Valley." *New York Times,* January 13, 2001, p. B9.

Markoff, John, and John Schwartz. "Many Tools of Big Brother Are Now Up and Running." *New York Times,* December 23, 2002, pp. C1, C4.

Martin, Douglas. "Charles Mount Dies at 60; Designed 300 Restaurants." *New York Times,* November 12, 2002, p. C21.

Martin, Steve. "Designer of Audio Packaging Enters Hell." *The New Yorker,* April 19, 1999, p. 53.

McCormick, Ernest J. *Human Factors Engineering.* 3d edition. New York: McGraw-Hill, 1970.

McGrane, Sally. "For a Seller of Innovation, a Bag of Technotricks." *New York Times,* February 11, 1999, p. G9.

McGuire, John M. "Stuck on Duct Tape." *St. Louis Post-Dispatch,* June 30, 1997, p. 1D.

McKibben, Gordon. *Cutting Edge: Gillette's Journey to Global Leadership.* Boston: Harvard Business School Press, 1998.

Miller, Donald L. *City of the Century: The Epic of Chicago and the Making of America.* New York: Simon & Schuster, 1996.

Miller, Ron. "To Boldly Paint What No Man Has Painted Before." *American Heritage of Invention and Technology,* Summer 2002, pp. 14–19.

Molotch, Harvey. *Where Stuff Comes From: How Toasters, Toilets, Computers, and Many Other Things Come to Be as They Are.* New York: Routledge, 2003.

Montfort, Nick. "Ten Passed Technologies." *Technology Review,* January/February 2001, pp. 90–93.

Morin, Monte. "Flashlight Feud Casts a Shadow." *Los Angeles Times,* August 13, 2002, p. 1.

Murphy, Kate. "Bigger Bar Code Inches Up on Retailers." *New York Times,* August 12, 2002, p. C3.

Muschamp, Herbert. "Fusing Beauty and Terror, Reverence and Desecration: The Fallen Concorde." *New York Times,* July 31, 2000, pp. E1, E3.

Norman, Donald A. *The Invisible Computer: Why Good Products Can Fail, the Personal Computer Is So Complex, and Information Appliances Are the Solution.* Cambridge, Mass: MIT Press, 1998.

———. *The Psychology of Everyday Things.* New York: Basic Books, 1988. (Reprinted in paperback as *The Design of Everyday Things.* New York: Doubleday, 1989.)

———. *Things That Make Us Smart: Defending Human Attributes in the Age of the Machine.* Reading, Mass.: Addison-Wesley, 1993.

———. *Turn Signals Are the Facial Expressions of Automobiles.* Reading, Mass.: Addison-Wesley, 1992.

Oehmke, Ted. "Cell Phones Ruin the Opera? Meet the Culprit." *New York Times,* January 6, 2000, pp. G1, G10.

O'Gorman, James F. "The Master Builder." *New York Times Book Review,* December 30, 2001, p. 9.

Palmeri, Christopher. "I Need to Be Making and Selling Things." *Forbes,* February 17, 1992, p. 97.

Panati, Charles. *Panati's Extraordinary Origins of Everyday Things.* New York: Harper & Row, 1987.

Parker, Ian. "Mr. Next." *The New Yorker,* January 13, 2003, pp. 29–30.

Paul, Lauren Gibbons. "Engineering News: Car Headlights Get on the Level." *Design News,* March 11, 2002, p. 54.

Pearlman, Chee. "A Little Brush, Reborn." *New York Times,* August 1, 2002, pp. F1, F4.

Peavey, Elizabeth. "Toll Taker for a Day." *Down East Magazine,* August 2002, pp. 86–87, 122–25.

Peterson, Iver. "E-ZPass Drivers in New Jersey to Pay New Fee." *New York Times,* July 12, 2002, p. B1.

Petroski, Henry. *The Book on the Bookshelf.* New York: Alfred A. Knopf, 1999.

———. *The Evolution of Useful Things.* New York: Alfred A. Knopf, 1992.

———. *Invention by Design: How Engineers Get from Thought to Thing.* Cambridge, Mass.: Harvard University Press, 1996.

———. *Remaking the World: Adventures in Engineering.* New York: Alfred A. Knopf, 1997.

———. *To Engineer Is Human: The Role of Failure in Successful Design.* New York: Vintage Books, 1992.

Pevenstein, Jack E. "Speculation on the Possibility of a Theory of Technology Evolution: Thoughts of an Engineer." Unpublished manuscript, National Institute of Standards and Technology, 2000.

Pinch, Trevor, and Frank Trocco. *Analog Days: The Invention and Impact of the Moog Synthesizer.* Cambridge, Mass.: Harvard University Press, 2002.

Poole, Robert. *Beyond Engineering: How Society Shapes Technology.* New York: Oxford University Press, 1997.

Prial, Frank J. "Now in the Best Bottles: Plastic." *New York Times,* August 8, 2001, pp. F1, F6.

Pridmore, Jay. *Inventive Genius: The History of the Museum of Science and Industry, Chicago.* Chicago: Museum of Science and Industry, 1996.

———. *Museum of Science and Industry, Chicago.* New York: Abrams, 1997.

Ramelli, Agostino. *The Various and Ingenious Machines of Agostino Ramelli (1588).* Translated and edited by Martha Teach Gnudi and Eugene S. Ferguson. Baltimore: Johns Hopkins University Press, 1976.

Rebora, Giovanni. *Culture of the Fork: A Brief History of Food in Europe.*

Translated by Albert Sonnenfeld. New York: Columbia University Press, 2001.

Savigny, J. *Treatise on the Use and Management of a Razor, with Practical Directions Relative to Its Appendages.* 2d edition. London: privately printed, n.d.

Schlager, Neil. *When Technology Fails: Significant Technological Disasters, Accidents, and Failures of the Twentieth Century.* Detroit: Gale Research, 1994.

Seabrook, John. "The Slow Lane." *The New Yorker,* September 2, 2002, pp. 120–29.

Selingo, Jeffrey. "It's the Cars, Not the Tires, That Squeal." *New York Times,* October 25, 2001, pp. D1, D8.

Sharke, Paul. "Let Light Be There." *Mechanical Engineering,* June 2001, pp. 70–73.

Simon, Herbert A. *The Sciences of the Artificial.* 2d edition. Cambridge, Mass.: MIT Press, 1981.

Smith, Roberta. "Just a Nonartist in the Art World, but Endlessly Seen and Cited." *New York Times,* January 21, 1998, pp. B1, B7.

Standage, Tom. *The Turk: The Life and Times of the Famous Eighteenth-Century Chess-Playing Machine.* New York: Walker, 2002.

Stanley, Autumn. *Mothers and Daughters of Invention: Notes for a Revised History of Technology.* Metuchen, N.J.: Scarecrow Press, 1993.

Steel, Ernest W. *Water Supply and Sewerage.* 4th edition. New York: McGraw-Hill, 1960.

Swearingen, David. "Brackett's New Market Open by Thanksgiving Weekend." [Maine] *Coastal Journal,* August 1, 2002, pp. 1, 4.

Templer, John. *The Staircase: History and Theories.* Cambridge, Mass.: MIT Press, 1992.

———. *The Staircase: Studies of Hazards, Falls, and Safer Design.* Cambridge, Mass.: MIT Press, 1992.

Tilley, Alvin R., and Henry Dreyfuss Associates. *The Measure of Man and Woman: Human Factors in Design.* New York: Whitney Library of Design, 1993.

Time-Life Books editors. *Inventive Genius.* Alexandria, Va.: Time-Life Books, 1991.

Tinniswood, Adrian. *His Invention So Fertile: A Life of Christopher Wren.* New York: Oxford University Press, 2001.

Tobin, James. *Great Projects: The Epic Story of the Building of America, from the Taming of the Mississippi to the Invention of the Internet.* New York: Free Press, 2001.

Toobin, Jeffrey. "Ashcroft's Ascent." *The New Yorker,* April 15, 2002, pp. 50–63.

Vanderbilt, Tom. "Brush with Greatness." *I. D. Magazine,* May 1999, pp. 64–71.

van Dulken, Stephen. *Inventing the 19th Century: 100 Inventions That Shaped the Victorian Age.* New York: New York University Press, 2001.

Vare, Ethlie Ann, and Greg Ptacek. *Patently Female: From AZT to TV Dinners, Stories of Women Inventors and Their Breakthrough Ideas.* New York: Wiley, 2002.

Venkataraman, S. T., and T. Iberall, eds. *Dextrous Robot Hands.* New York: Springer-Verlag, 1990.

Vincenti, Walter. *What Engineers Know and How They Know It: Analytical Studies from Aeronautical History.* Baltimore: Johns Hopkins University Press, 1990.

Vitruvius. *The Ten Books on Architecture.* Translated by Morris Hicky Morgan. New York: Dover Publications, 1960.

Vogel, Steven. *Cats' Paws and Catapults: Mechanical Worlds of Nature and People.* New York: Norton, 1998.

———. *Prime Mover: The Muscle-Powered World.* New York: Norton, 2001.

Voss-Hubbard, Anke. "Hugh Moore and the Selling of the Paper Cup: A History of the Dixie Cup Company, 1907–1957." *Canal History and Technology Proceedings XV,* March 9, 1996, pp. 83–101.

Wade, Betsy. "Zapping Tolls, with Asterisks." *New York Times,* November 5, 2000, Section 5, p. 4.

Ward, Leah Beth. "At Nike, Function over Fashion." *New York Times,* April 28, 2002, Section 3, p. 2.

Watson, Bruce. *The Man Who Changed How Boys and Toys Were Made.* New York: Viking, 2002.

Watson, J. G. *A Short History: The Institution of Civil Engineers.* London: Thomas Telford, 1982.

Webb, Pauline, and Mark Suggitt. *Gadgets and Necessities: An Encyclopedia of Household Innovations.* Santa Barbara, Calif.: ABC-CLIO, 2000.

Wikell, Goran. "The Layout of Digits on Pushbutton Telephones." *Tele,* no. 1 (1982): 34–45.

Wilson, Joe. *Ductigami: The Art of the Tape.* Erin, Ontario: Boston Mills Press, 1999.

Wohleber, Curt. "The Remote Control." *American Heritage of Invention and Technology,* Winter 2001, pp. 6–7.

Wulf, William. "The Urgency of Engineering Education Reform." American

Society for Engineering Education, Annual Conference and Exposition, Montreal, June 16–19, 2002, http://www.asee.org/conferences/annual2002/wulfplenary.cfm.

Wulf, William A., and George M. C. Fisher. "A Makeover for Engineering Education." *Engineering Times,* August/September 2002, p. 5.

Zigman, Laura. "Good Fences Make Big Bills." *New York Times,* August 29, 2002, p. F10.

ILLUSTRATION CREDITS

6 Plastic tripods. *Digital photo by Catherine Petroski.*
18 Text distorted through glass. *Digital photo by Catherine Petroski.*
26 Image in bubble. *Digital photo by Catherine Petroski.*
31 Early paper cup. *Dixie Cup Collection, David Bishop Skillman Library, Lafayette College. Used with permission.*
33 Luellen Cup and Water Vendor. *Dixie Cup Collection, David Bishop Skillman Library, Lafayette College. Used with permission.*
42 Brita pitcher. *Digital photo by Catherine Petroski.*
63 Volvo cup holder. *Digital photo by Catherine Petroski.*
104 Knight's first paper-bag machine. *U.S. Patent No. 116,842.*
105 Folding and pasting steps for making paper bag. *U.S. Patent No. 116,842.*
107 Crowell's patent for paper bag. *U.S. Patent No. 123,811.*
115 Goldman's shopping cart patent. *U.S. Patent No. 2,155,896.*
139 WD-40 spray can. *Digital photo by Catherine Petroski.*
144 Moen shower control. *Digital photo by Catherine Petroski.*
152 Vegetable peelers. *Digital photo by Catherine Petroski.*
159 Steelcase office chair. *Digital image courtesy of Steelcase Inc.*
160 Aeron office chair. *Image (from transparency) courtesy of Herman Miller.*
169 Measures for design. *From* The Measure of Man and Woman.
172 Doorknob and light switch. *Digital photo by Catherine Petroski.*
182 Number pads. *Digital photo by Catherine Petroski.*
196 Tech Box. *Digital image courtesy of IDEO. Photo by Joe Watson.*
214 Oral-B toothbrushes. *Digital photo by Catherine Petroski.*

INDEX

Page numbers in *italics* refer to illustrations and their captions.